AF774435

BioMethods Vol. 8

Series Editors

Dr. T. Meier
Physiologisches Institut
Universität Basel
Vesalgasse 1
CH-4051 Basel
Switzerland

Dr. H.-P. Saluz
Hans-Knöll-Institut
für Naturstofforschung e.V.
Beutenbergstr. 11
D-07745 Jena
Germany

I. Cosic

The Resonant Recognition Model of Macromolecular Bioactivity

Theory and Applications

Birkhäuser Verlag
Basel · Boston · Berlin

Author

Dr. Irena Cosic
Department of Electrical and
Computer Systems Engineering
Monash University, Caulfield Campus
900 Dandenong Rd.
Caulfield East, VIC 3145
Australia

Library of Congress Cataloging-in-Publication Data

Cosic, I. (Irena), 1952–
The resonant recognition model of macromolecular bioactivity : theory and applications / I. Cosic.
p. cm. -- (BioMethods : Vol. 8)
Includes bibliographical references and index.

1. Protein binding. 2. DNA-protein interactions. 3. Proteins-Spectra. 4. Proteins--Electric properties.
5. Bioelectrochemistry. I. Title. II. Series
QP551.C727 1997
5 72.8'4--dc21

Deutsche Bibliothek Cataloging-in-Publication Data

Cosic, Irena :
The resonant recognition model of macromolecular bioactivity : theory and applications / I. Cosic. - Basel ; Boston ; Berlin: Birkhäuser, 1997
(BioMethods ; Vol. 8)

ISBN 978-3-0348-7477-9 ISBN 978-3-0348-7475-5 (eBook)

DOI 10.1007/978-3-0348-7475-5

NE: GT

The publisher and editor can give no guarantee for the information on drug dosage and administration contained in this publication. The respective user must check its accuracy by consulting other sources of reference in each individual case.

The use of registered names, trademarks etc. in this publication, even if not identified as such, does not imply that they are exempt from the relevant protective laws and regulations or free for general use.

This work is subject to copyright. All rights are reserved, whether the whole or part of the material is concerned, specifically the rights of translation, reprinting, re-use of illustrations, recitation, broadcasting, reproduction on microfilms or in other ways, and storage in data banks. For any kind of use permission of the copyright owner must be obtained.

© 1997 Birkhäuser Verlag, PO Box 133, CH-4010 Basel, Switzerland
Printed on acid-free paper produced from chlorine-free pulp. TCF ∞

Softcover reprint of the hardcover 1st edition 1997

9 8 7 6 5 4 3 2 1

"What is that causes inorganic matter to run into organic forms? It is the sun's heat and light."

Tesla, 1915

To my husband Drasko, daughter Katarina and son Neil

Contents

Acknowledgements

I would like to thank Associate Professor Paul Voumard for reading and reviewing this book. I would also like to thank my collaborators, assistants and students who helped me over 15 years of research with the RRM. Funding by Monash University and the Australian Research Council is also acknowledged.

Summary

Biological processes in any living organism are based on selective interactions between particular biomolecules. In most cases, these interactions involve and are driven by proteins, which are the main conductors of any life process within the organism. The physical nature of these interactions is still not well known.

This book presents an entirely new approach to analysis of biomolecular interactions, in particular protein-protein and protein-DNA interactions, based on the assumption that these interactions are electromagnetic in nature. This new approach is the basis of the Resonant Recognition Model (RRM), which was developed over the last 15 years.

Certain periodicities within the distribution of energies of delocalised electrons along a protein molecule are crucial to the protein's biological function, i.e. interaction with its target. If protein conductivity were introduced, then charges moving through the protein backbone might produce electromagnetic irradiation or absorption with spectral characteristics corresponding to energy distribution along the protein. The RRM is capable of calculating these spectral characteristics, which we hypothesized would be in the range of the infrared and visible light. These characteristics were confirmed with frequency characteristics obtained experimentally for certain light-induced biological processes.

This new concept can be applied to the following:

- the definition of protein or DNA functions;
- the definition of protein or DNA targets and the analysis of their mutual recognition;
- the prediction of amino acids in the protein or nucleotides in the DNA which are essential to the function of the macromolecules;
- the prediction of functionally relevant mutations in proteins and/or DNA;
- the design of completely new peptides or DNA fragments with desired spectral characteristics and, consequently, desired corresponding biological activities.

All these applications are presented in a number of examples that include some already tested in biological systems, for example fibroblast growth factor analogues and human immunodeficiency virus envelope-mimicking peptides designed de novo using the RRM.

1 Introduction

1.1 Structure-function relationships in proteins and DNA: definition of the problem

Biological processes in any living organism are based on selective interactions between particular biomolecules. In most cases, these interactions involve and are driven by proteins and/or DNA, which are the main informational contributors in any living process within the organism. Proteins and DNA are linear macromolecules composed of sequentially linked constitutive elements, namely amino acids or nucleotides. Proteins, but DNA as well, can only express their biological function when they achieve a certain active three-dimensional (3D) structure. Their biological function, as well as their active (3D) structure, is determined primarily by the amino acid or nucleotide sequence within the macromolecule. Although both function and structure of a number of protein and DNA sequences are known [1], the crucial problem of understanding how the biological function and active 3D structure are "written" within these sequences still remains unresolved. If biological function is considered as a selective interaction of a protein and its target, then a more fundamental question arises, namely, What is the physical basis of this interaction and how is selectivity achieved? Once this understanding has been gained, it should be possible to design peptides, proteins and DNA sequences *de novo*, with a chosen, desired biological function, and thus to produce new and more effective drugs and other biotechnological products.

 ©1997, Birkhäuser Verlag Basel

1.2 Existing approaches and their limitations

This chapter describes existing approaches to the analysis of protein and DNA structure-function relationships.

Primary, secondary and tertiary structure analysis

There have been many attempts to discover the rules governing the coding of the biological function into the sequence of amino acids within the protein [2-12]. Typical approaches deal with either the homology characterisation of specific features of the primary and secondary structures of proteins or with molecular modelling of protein tertiary structure. Although such approaches enable significant insight into protein structure and active site location, they still do not provide sufficient knowledge about the informational, structural and physicochemical parameters that are crucial to the selectivity of protein interactions that can be used for *de novo* design of peptide or protein analogues with the desired biological activity. Current knowledge in the field of computer-aided molecular modelling and protein structure-function analysis can be classified in terms of the primary, secondary and tertiary structure analysis of proteins.

The RRM essentially belongs to the field of protein and DNA sequence analysis. Research approaches in this field can be classified into two groups:

1. mathematical analysis of amino acid or nucleotide arrangement aimed at searching for information on biological function;
2. analysis of physical processes inside the macromolecule, which could be relevant to its biological function.

The most commonly used method from the first group is homology searching. The main idea of this method is that sequences with the same biological functions do have sequence homology and that these homologous fragments carry the main information about function. This method is very useful in the case of conserved sequences (i.e. histones, haemoglobins, insulins etc.) or of some DNA regulatory sequences, where the existence of a specific core sequence of

5-7 nucleotides has been established. For promoters, this is the TATA box [13] and for enhancers, the GTGG–GT box [14]. However, there are many of examples of sequences which have the same biological activity but do not have a significant degree of homology, as well as examples where a significant degree of homology exists between functionally unrelated sequences. A general opinion is that homology denotes the same ancester molecule but not a similar function [3].

The problem of protein structure-function relationships and the definition of protein active sites is far from solved. Numerous costs and time-consuming experimental procedures (e.g. bioassays, cell cultures, mutagenesis research and in vivo experiments) are involved in testing the functionality of different fragments or even of a single amino acid within a protein. And a number of computational modelling procedures are required to predict either conserved regions within the protein sequence that suggest an active site or a protein folded structure that suggests the minimum energy conformation [5-12]. Although conserved regions are known and active conformations have been successfully predicted or identified by crystallographic or solution measurements, it is still very difficult to position an active site within the 3D protein structure without additional biological experimental results. In most cases, an active site is built up of amino acid residues which are not sequentially linked, although they can form a continuous interactive surface in the 3D protein structure.

Typically, the search for homologies among amino acid sequences is used to identify functionally relevant amino acids. The main concept behind this method is that proteins with the same biological function share amino acid sequence alignments, and these homologous fragments carry the main information about protein function [1, 3]. A problem arises in cases of very conserved proteins, where it is difficult to distinguish functionally important amino acids from other homologous ones [3]. Additional problems exist with proteins where the active site is not formed uniquely of conserved amino acids and with cases where very conserved protein sequences do not even have the same biological function. Optimal alignment programs [2] with new improvements in sequence analysis are designed to distinguish between analogous and homologous sequences, but they are still based on sequence similarities. Problems with optimal alignment programs include difficulties associated with the length of the sequence string and the insertion of gaps in order to increase the number of matching residues. Inserting gaps too liberally and assigning gap-weights arbitrarily can lead to biologically irrelevant alignments. Generally, the problem with sequence alignment algorithms lies in their search for local homologies rather than for homologies of properties distributed along the protein.

There are other methods of sequence analysis, primarily statistical, and most of them are presented in the University of Wisconsin software package (GCG software package) [2]. Generally, they are all based on a multiple sequence alignment, with or without gaps. Furthermore, the matching score is calculated in a number of different ways, indicating conserved regions. If sequences compared have some biological characteristic in common, then conserved regions are proposed as important regions for this characteristic. There are also a number of methods and computer programs for sequence alignment and identification of conserved regions, as well as for searching for particular patterns along the protein or DNA sequence [2, 13, 14].

The hydropathy method of Dr. R. Doolittle [3] is a sequence analysis which belongs essentially to the first group of methods but has elements of a physical approach as well. The main idea of this approach is to represent a protein sequence as a numerical sequence by representing each amino acid with a corresponding hydropathy value. After smoothing this curve, it is possible to see the distribution of hydropathy along the sequence. This is important for estimating the active site of the protein.

The methods of sequence analysis, based on physical processes in the macromolecule, may be divided into two groups: analysis of structure, and analysis of charge transfer through the backbone of the macromolecule.

The analysis of structure incorporates a secondary structure (2D) and tertiary (3D) analysis and predictions. The 2D structure of proteins relates to analysis and prediction of whether each residue in the protein forms part of an α helix (α), a β sheet (β) or is in a turn formation. This is usually analysed by statistical observation of the probabilities that one amino acid, in a particular neighbourhood, is in a particular secondary structure. There are a number of different approaches to 2D protein structure analysis. In most of these calculations, certain amino acid parameters have been used (e.g. hydrophobicity, hydropathy etc.) as parameters to define neighbouring conditions. These methods can predict 2D protein structure with different accuracies ranging from 50 to 70% [15].

The tertiary structure is obtained experimentally by nuclear magnetic resonance (NMR) spectroscopy or X-ray crystallography. A number of attempts were made to predict tertiary structures theoretically, but these methods have limited success and will be discussed later. Nevertheless, the whole tertiary structure, and what is more important, the biological function are in some way written in the primary structure. Thus, if one could read the biological function from the primary structure, one does not need to know its tertiary structure.

The analysis of protein and DNA sequences, which is the most interesting for my work, involves the investigation of charge transfer through the backbone of macromolecules. Early results of J. Ladik [16] show that "the valency and conducting bandwidth are found to be large enough to make electronic delocalisation through DNA backbone possible. Different mechanisms for charge carrier transport in DNA are compared on the basis of the electron and hole effective mass. Conducting along the backbone seems to be at least as probable as through the aperiodic system of the superimposed nucleotide bases." In other words, there is enough energy of delocalised electrons in the DNA backbone to make charge transfer possible, but the mechanisms of this transfer are still not known.

W. A. Little [17] analysed the possibility of superconductive effect in biopolymers, showing that superconductive states, at around room temperature, should be distinguished by certain unique chemical properties which could have considerable biological significance.

Davidov and his co-workers [18, 19] have considered the idea that α-helical proteins may facilitate electron transfer through a soliton mechanism. In this case, an extra electron causes a lattice distortion in the protein which stabilises the electron motion. There is also some experimental evidence of soliton existence in DNA and proteins [20]. Models of charge transfer through the protein and DNA backbone are concepts in macromolecule dynamics which merit the careful attention of biologists.

Three-dimensional protein structure: importance of function

The folding of the linear, primary polymer chain of a protein into a defined 3D structure results in a spatial relationship between the various constituent amino acids that is crucial for determining the functional behaviour of the protein. In particular, the widely accepted model of protein interactions proposes that the selectivity of these interactions is based on the structural matching between active sites of interacting molecules.

Since the first low resolution X-ray structure of globular protein was obtained in 1958 by Kendrew et al. [21], hundreds of other 3D protein structures were identified using X-ray and NMR techniques. The Brookhaven protein structures (PDB) database contains atomic coordinates for over 3000 proteins, and new entries are constantly being added.

The folding of the linear, primary sequence of a polypeptide or protein into a defined 3D structure results in complex spatial relationships between the various constituent amino acids. Inherent in this folding process is the topographical positioning of certain key amino acids which are crucial for determining the functional behaviour of the whole protein. Over the past two decades substantial scientific effort has been directed towards addressing a variety of fundamental issues related to protein folding. In particular, considerable attention has been focused on the essential paradigmatic question pertinent to the nature of the relationships between the protein hierarchical structure and protein biological function, namely, which amino acids within the primary sequence dictate the processes of biorecognition associated with the orientation of the substrate, in the case of enzymes, and aprpriate "docking", in the case of hormone-receptor interactions or antigen-antibody interactions.

Experimentally, the tertiary (3D) structure and stability of polypeptides and proteins have been studied using a great variety of physical techniques, including X-ray crystallography, circular dichroism, fluorescence spectroscopy and, more recently, 2D NMR [9]. However, the availability in recent years of enormously increased databases on protein primary sequences has reinforced the need for theoretical methods capable of analysing protein structure-function relationships. All previous theoretical approaches can be placed into the following two categories based on the level of the structural hierarchy analysed: (1) approaches which make secondary and tertiary structure predictions from known primary structure and (2) approaches which seek a direct correlation between primary structure and function of proteins.

The protein native conformation is unique, and the same globular conformation is obtained on successive refolding, although there are small fluctuations around the native state [22]. Furthermore, the conformational transition appears to be well fitted into the two-state model, that is, the most probable states are either a native or a denatured structure [22]. Surprisingly, the energy difference between these two states is very small, in the order of several hundred calories per mole of residues. The following question then arises: Is the native structure at the global free energy minimum, or does it result from kinetic trapping [22]? It is widely accepted that a native structure represents the structure of minimum energy, although this has yet to be proved. A random search of all conformations of a protein having even 10^{23} conformations would take about 10^5 years. Thus, having in mind the speed of native folding, whatever the folding mechanism is, it is not random. The situation has become even more complicated since chaperones, the proteins that help other proteins to fold, were discovered. There are a number of different computational approaches for analysis and prediction of protein 3D structure. However, none of them has solved the problem completely.

Three-dimensional protein structure: predictions

The tertiary structure and stability of proteins have been experimentally studied using such techniques as X-ray crystallography, circular dichroism, fluorescence spectroscopy and 2D NMR. However, these methods have been limited due to the need for relatively large amounts of protein and the inability of many techniques to detect low abundance of conformational intermediates. The methods may also be limited by structural distortions caused by particular techniques, such as crystallisation. The increasing database of experimentally derived protein primary structures combined with computer algorithms for performing molecular mechanics and dynamics has the potential to establish computational algorithms as a powerful tool to study protein tertiary structure and predict peptide/protein active conformations. These methods do not enable the prediction of 3D protein structure solely from its sequence, and thus they tend to use a number of constraints from experimental measurements, or they are based on the sequential and functional homology of proteins with known 3D structures.

Since the first low-resolution X-ray structure of globular protein was obtained, the ability to predict native protein, biologically active 3D structures from the sequence of amino acids has been a long-sought objective of theoretical biophysical chemistry [22]. Theorists are faced with two interrelated problems: (1) identifying the most suitable potential energy function, and (2) defining practical algorithms for predicting structure.

There are a number of different approaches to predicting 3D protein structure. The easiest is to predict the structure of various small constrained systems, such as hypervariable loops in antibodies, where a full search of conformational space is possible. Similar procedures may be applied to larger systems. However, the possibility of identifying global minima is exponentially less as a system is larger. In these cases, the minimisation usually ends up in one of various local minima [10], and the final structure depends largely on the initial structure used for the minimisation. Thermal energy fluctuations were introduced to the system in order to force a molecule to jump over energy barriers and eventually to find global minima. These molecular dynamic simulations are based on solving the Newtonian equations of motion for the atoms of the protein and any surrounding solvent [12]. These methods have not yet been successful with large molecules because they require many calculations and because minimisation can be "trapped" in one of a number of local minima. To investigate conformational space more efficiently, a statistical Monte Carlo method is used; to overcome the multiple minima problem, a hybrid Monte Carlo/energy minimisation procedure was proposed. However, none of these approaches has been successful in predicting the 3D structure of proteins in general.

A methodology related in concept to a full conformational search is the build-up procedure developed by Scheraga and co-workers [23]. This is based on the assumption that the local intraresidue interactions are the dominant factor in determining the conformation of a protein. The procedure starts with low-energy conformations of dipeptides, then combines them to find low-energy conformations of tripeptides and so on. Although this method is very efficient for small peptides, it could not be succesfully applied to larger proteins.

Even if we were able to predict the 3D structure of protein, the question would remain, is it sufficient to know the 3D structure in order to understand its function and consequently to be able to design, *de novo*, functionally active proteins? The fact that interacting molecules fit structurally at the positions of active sites does not exclude the possibility that many other structures can also fit each other but not interact. In addition, because these parameters vary from the case to case, there is no definite measure of how well interacting structures have to fit each other to be able to interact and what the critical docking energy for these interactions is. Thus, we cannot conclude that structure is the only and critical parameter for selective recognition between interacting macromolecules (proteins, DNA, RNA etc.). This book represents a novel approach to macromolecular interactions by introducing a new and very selective parameter, resonant energy transfer between interacting macromolecules, as a critical parameter in macromolecular interactions (functions).

A variety of empirical algorithms are now available for predicting the secondary structure of polypeptides and proteins from knowledge of their primary sequence

[5-7]. These predictions (e.g. as Ramachandran diagrams, Chou-Fasman plots etc.) are often presented in the form of preferred regions of regular secondary folding (i.e. α helix, β sheet etc.) or predicted tertiary packing of 2D structures (e.g. Finer-Moore packing models) for a given peptide chain. While these approaches have been relatively successful in predicting secondary structure (with probabilities in the region of 60-70% with selected examples) [6], characterisation of the complete 3D structure and determination of the direct relationship to protein function has yet to be achieved via these methods [8].

1.3 Novelty of the RRM

The RRM [24, 26] is a physical and mathematical model which interprets protein sequence linear information using methods of signal analysis. The model presented describes a new view of biologically relevant intermolecular interactions, which, it is suggested, consist in the transfer of resonant energy between molecules. This model could have a great impact on molecular biology and consequently on medicine, pharmacology and agriculture.

1.4. Applications of the RRM

Once an understanding of the nature of protein interactions and their selectivity has been reached, this knowledge can be used to benefit molecular biology and consequently medicine, pharmacology and agriculture. In general, possibilities emerge, (1) to define a particular function of a protein or DNA fragment; (2) to predict functionally important amino acids within the protein sequence and thus to propose effective mutations; (3) to analyse and predict the possibility of macromolecular interactions in particular protein-DNA interactions; and (4) to design sequences with desired spectral and, consequently, functional characteristics. These applications of the RRM are described here in a number of examples.

2 RRM Basic Concepts

The RRM is based on the premise that proteins with the same biological function or interactive activity have the same periodic components in the distribution of delocalised electron energies along the protein molecule. This has been shown by considering protein as a numerical sequence representing each amino acid in the sequence with a corresponding electron-ion interaction value. This numerical sequence then represents distribution of delocalised electron energies along the molecule. Numerical series obtained in this way can then be transformed into frequency spectra using discrete Fourier transform (DFT) techniques.

The new physico-mathematical approach presented here is called the Resonant Recognition Model [24-35]. The RRM is based on the representation of the protein primary structure as a numerical series by assigning to each amino acid a physical parameter value relevant to the protein's biological activity. Although a number of amino acid indices (222 have been published up to now [36, 37]) have been found to correlate in some ways with the biological activity of the whole protein, our investigations [24-35], as well as studies of other authors [38-40], have shown that optimum correlation can be achieved with parameters which are related to the energy of the delocalised electrons of each amino acid. These findings can be explained by the fact that the electrons delocalised in the particular amino acid have the strongest impact on the electronic distribution of the whole protein. In our studies, the energy of delocalised electrons (calculated as the electron-ion interaction pseudopotential, EIIP [41, 42]) of each amino acid residue was used. The resulting numerical series represents the distribution of the free electron energies along the protein. This numerical series was converted into a discrete Fourier spectrum which carried the same information about the arrangement of amino acids in the sequence as did the original numerical sequence [43]. The initial step in the RRM, where protein sequences are represented as numerical spectra, was called the In-

 ©1997, Birkhäuser Verlag Basel

formational Spectrum Method (ISM) [27-29]. Later, when it was found that proteins can recognise their targets on the basis of the same characteristic frequency [24], which is in fact resonant recognition, the whole model was renamed more appropriately as the Resonant Recognition Model.

Approaches similar to the RRM, based on the Fourier transform and physical characteristics of amino acids, have been successfully applied by Mandell, who has shown that the characteristic hydrophobic mass energy Fourier modes are signatures of isomorphism and immunological reactivates [44]. Viari and co-workers have used our RRM approach with scale- independent coding to localise biologically relevant patterns in calcium-binding proteins [45].

2.1 Definition of macromolecular functional characteristics: characteristic frequency

To determine the common components of a group of proteins with the same biological function, the multiple cross-spectral function was used. Peaks in this function denote common frequency components for the sequences analysed. The presence of a significant peak in a multiple cross-spectral function of a group of proteins with the same biological function means that all of the analysed sequences within the group have this frequency in common. It has been shown with more than 1000 proteins from more than 30 functional groups that each biological function is characterised by one unique frequency. Furthermore, it has been shown that proteins and their receptors do have the same characteristic frequen-cies but with opposite phases at these frequencies. This should imply that these mathematically obtained characteristic frequencies are an important parameter of macromolecular biological function and interaction. Furthermore, biological interactions between proteins and their targets can be observed as a resonant process.

The RRM is a physical and mathematical model which interprets protein sequence linear information using methods of signal analysis. It comprises two stages. The first involves the transformation of the amino acid sequence into a numerical sequence. Each amino acid is represented by the value of the EIIP [16], which describes the average energy states of all valence electrons in a particular amino acid.

The EIIP values for each amino acid were calculated using the following general model pseudopotential [29].

$$<\vec{k} + q \mid w \mid \vec{k}> = 0.25\, Z \sin(\pi\, 1.04\, Z)/(2\pi)$$

where q is a change of momentum of the delocalised electron in the interaction with potential w, while

$$Z = (\Sigma Z_i)/N$$

where Z_i is the number of valence electrons of the i-th component of each amino acid and N is the total number of atoms in the amino acid. The EIIP values for 20 amino acids as well as for 5 nucleotides (the whole procedure can be applied to DNA and RNA, too) are shown in Table 2.1. Each amino acid or nucleotide, irrespective of its position in a sequence, can thus be represented by a unique number.

Nucleotide	EIIP (Ry)
A	0.1260
G	0.0806
T	0.1335
C	0.1340
U	0.0289

Amino acid	EIIP (Ry)
Leu	0.0000
Ile	0.0000
Asn	0.0036
Gly	0.0050
Val	0.0057
Glu	0.0058
Pro	0.0198
His	0.0242
Lys	0.0371
Ala	0.0373
Tyr	0.0516
Trp	0.0548
Gln	0.0761
Met	0.0823
Ser	0.0829
Cys	0.0829
Thr	0.0941
Phe	0.0946
Arg	0.0959
Asp	0.1263

Table 2.1 The electron-ion interaction potential (EIIP) values for nucleotides and amino acids

Numerical series obtained this way are then analysed by digital signal analysis methods in order to extract information pertinent to the biological function. Any numerical signal of length N can be represented as a sum of N/2 sinusoids, each characterised by a corresponding amplitude A, frequency f and phase φ (Fig. 2.1). This can then be regarded as a frequency spectrum which is a function of amplitude versus corresponding frequency (amplitude spectrum) and phase versus frequency (phase spectrum) for these sinusoids. Amplitude and phase spectra contain the same information as the original signal, but this information is represented in the frequency domain rather than in the space (or time) domain.

Figure 2.1

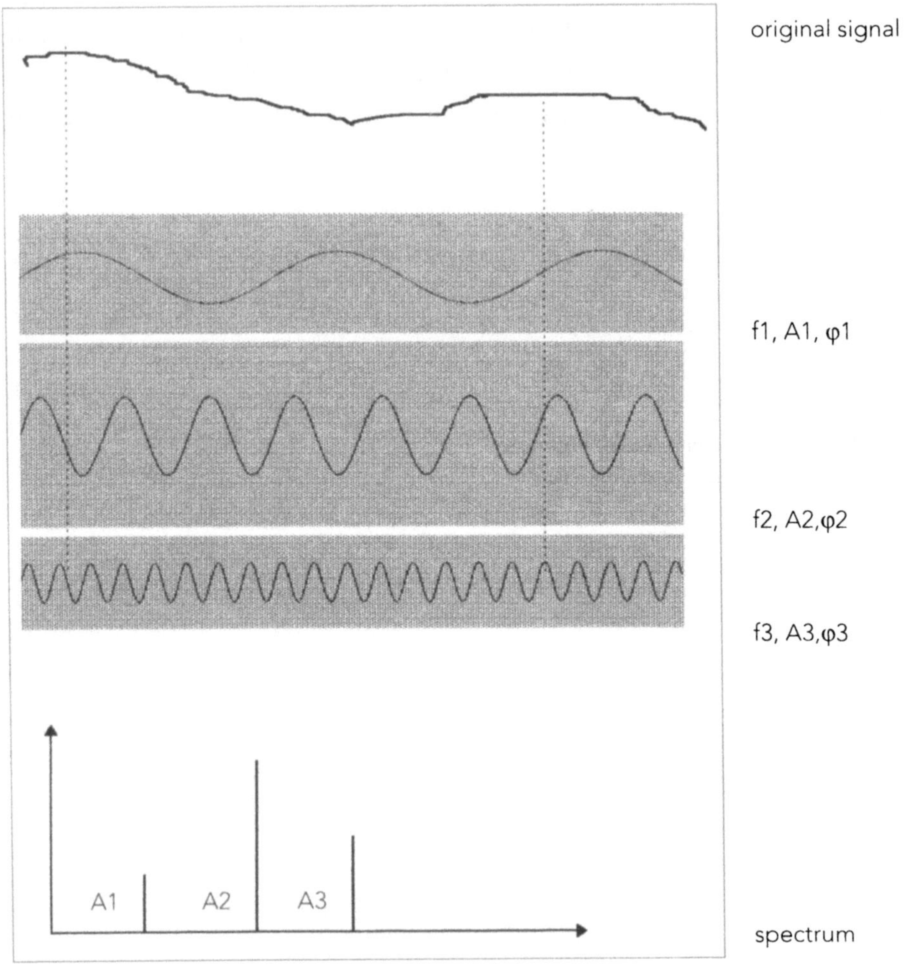

The original numerical sequence is transformed into the frequency domain using DFT. DFT coefficients X(n) are defined as

$$X(n) = \sum x(m)\, e^{-j(2\pi/N)nm} \quad n = 1,2, \ldots, N/2$$

where x(m) is the m-th member of the original numerical series, N is the total number of points in this series and X(n) is the DFT coefficient. These coefficients describe the amplitude, phase and frequency of sinusoids of the original signal. The absolute values of complex DFT coefficients define the amplitude spectrum, while their phases define the phase spectrum as follows:

$$X(n) = |X(n)|\, e^{-j\varphi(n)}$$

where |X(n)| represents the amplitude spectrum and φ(n) represents the phase spectrum. Complete information about the original sequence is contained in both spectral functions. However, in some cases, relevant information can be contained only in one of these two spectral functions. For example, information relevant to understanding speech is contained only in the amplitude spectrum of sound. In such cases, it is practical to analyse the energy density spectrum, which is defined as follows:

$$S(n) = X(n)AX(n)^* = |X(n)|^2 \quad n = 1,2, \ldots, N/2$$

Information contained in the energy density spectrum was initially analysed in the case of proteins and DNA sequences. This spectrum is denoted as "information spectrum" (IS).

In this procedure, macromolecular sequences are analysed as discrete signals. As the average distance between amino acid residues in a polypeptide chain is about 3.8 Å, it can be assumed that the points in the derived numerical sequence are

◀ ***Figure 2.1 The frequency spectrum is a function of the amplitude of the signal versus frequency (amplitude spectrum) and phase versus frequency (phase spectrum)***

Both amplitude and phase spectra contain the same information as the original signal, but this information is expressed in the other domain. Any time-changing discrete signal can be represented as the sum of a limited number of sinusoids where each is characterised by a corresponding amplitude A, frequency f and phase φ, thus another function can then be represented with frequency on the x-axis and amplitude or phase of the corresponding sinusoids on the y-axis.

equidistant. For further numerical analysis, the distance between points in these numerical sequences is set at an arbitrary value d = 1. Then the maximum frequency in the spectrum is F = 1/2d = 0.5. The total number of points in the sequence influences the resolution of the spectrum only. Thus, for an N-point sequence the resolution in the spectrum is equal to 1/N. The n-th point in the spectral function corresponds to the frequency f = n/N.

In order to extract common spectral characteristics from sequences having the same or similar biological functions, the following cross-spectral function was used:

$$Sn = Xn\,Yn^{*} \quad n = 1,2,\ldots,N/2$$

where Xn are the DFT coefficients of the series x(m) and Yn^{*} are complex conjugate DFT coefficients of the series y(m). Peak frequencies in the amplitude cross-spectral function define the common frequency components of the two sequences analysed. The whole procedure can be mapped as protein sequence → numerical series → amplitude spectra → cross-spectra. This is represented in Figure 2.2 using the example of hemoglobins.

To determine the common frequency components for a group of protein sequences, we have calculated the absolute values of the multiple cross-spectral function coefficients M, which are defined as follows:

$$|Mn| = |X1n|A|X2n| \ldots |XMn| \quad n = 1,2,\ldots,N/2$$

Peak frequencies in such a multiple cross-spectral function denote common frequency components for all sequences analysed. Signal-to-noise ratio (S/N) for each peak is defined as a measure of similarity between sequences analysed. S/N is calculated as the ratio between signal intensity at the particular peak frequency and the mean value over the whole spectrum. The extensive experience gained from pre-

Figure 2.2 The RRM procedure

(a) Sequences of α and β human hemoglobins, (b) graphical representation of the corresponding numerical sequences obtained by replacing every amino acid with its EIIP value; (c) spectra of both α and β human hemoglobins; (d) cross-spectral function of the spectra presented in (c). The prominent peaks denote common frequency components. The abscissa represents RRM frequencies, and the ordinate is the normalised intensity. ▶

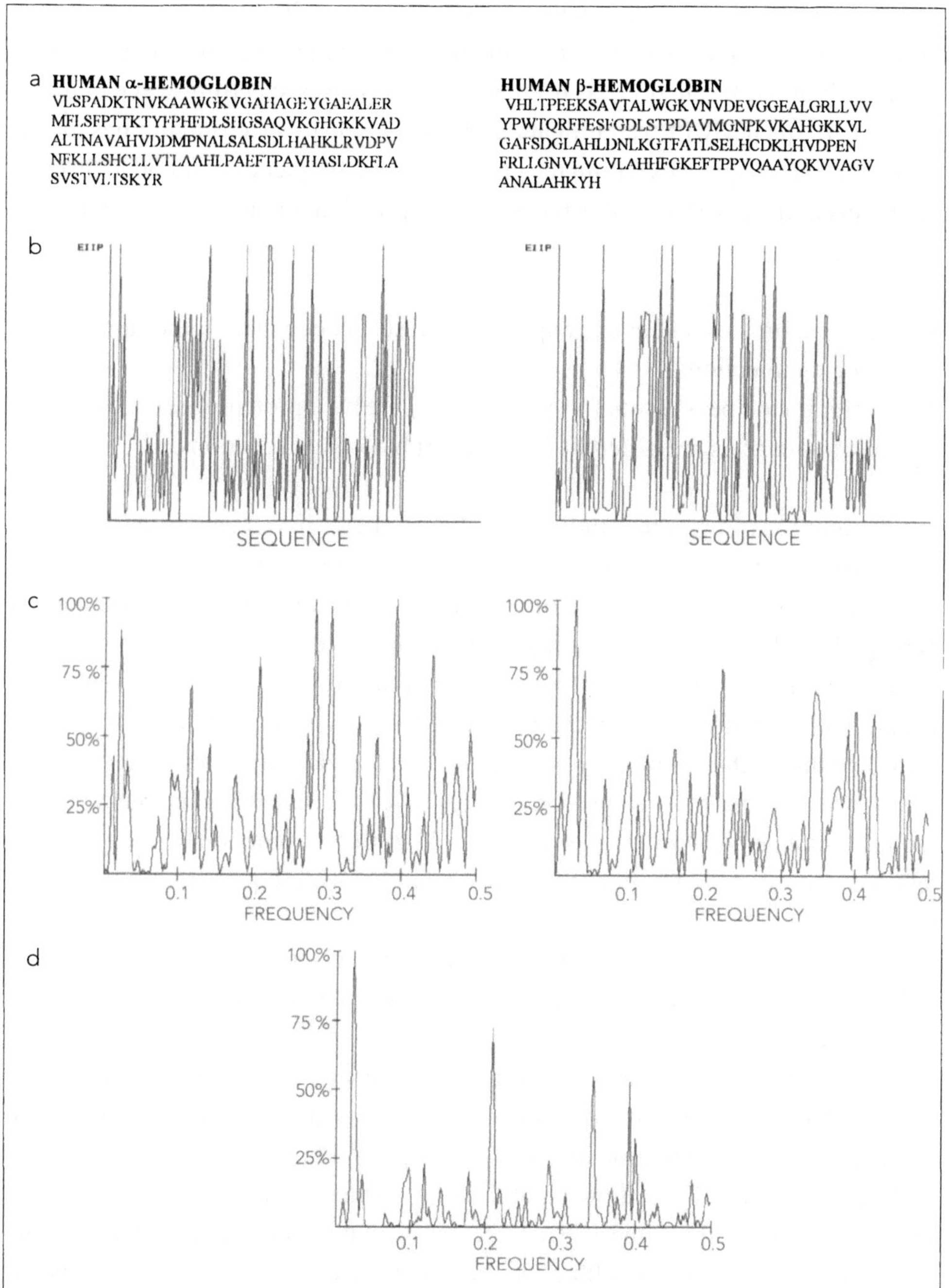
a
HUMAN α-HEMOGLOBIN
VLSPADKTNVKAAWGKVGAHAGEYGAEALER
MFLSFPTTKTYFPHFDLSHGSAQVKGHGKKVAD
ALTNAVAHVDDMPNALSALSDLHAHKLRVDPV
NFKLLSHCLLVTLAAHLPAEFTPAVHASLDKFLA
SVSTVLTSKYR
HUMAN β-HEMOGLOBIN
VHLTPEEKSAVTALWGKVNVDEVGGEALGRLLVV
YPWTQRFFESFGDLSTPDAVMGNPKVKAHGKKVL
GAFSDGLAHLDNLKGTFATLSELHCDKLHVDPEN
FRLLGNVLVCVLAHHFGKEFTPPVQAAYQKVVAGV
ANALAHKYH
b
EIIP
SEQUENCE
EIIP
SEQUENCE
c
100%
75 %
50%
25%
0.1
0.2
0.3
0.4
0.5
FREQUENCY
100%
75 %
50%
25%
0.1
0.2
0.3
0.4
0.5
FREQUENCY
d
100%
75 %
50%
25%
0.1
0.2
0.3
0.4
0.5
FREQUENCY

Figure 2.2

vious research [24-30] suggests that an S/N of at least 20 can be considered significant. The multiple cross-spectral function for a large group of sequences with the same biological function has been named "consensus spectrum". The presence of a peak frequency with a significant S/N in a consensus spectrum implies that all of the analysed sequences within the group have one frequency component in common. This frequency is related to the biological function provided the following criteria are met:

1. One peak only exists for a group of protein sequences sharing the same biological function.
2. No significant peak exists for biologically unrelated protein sequences.
3. Peak frequencies are different for different biological functions.

In a number of previous studies, the above criteria have been tested with over 1000 proteins from over 30 functional groups [24-35]. Figure 2.3 represents the evaluation of the consensus spectrum of one functional group of proteins, namely hemoglobins. It is evident that a single prominent peak describes the only frequency component that is common to the whole group of proteins. Figure 2.4 represents the evaluation of the cross-spectral function of biologically unrelated sequences. It is evident that such a multiple cross-spectral function does not contain any relevant peaks, which means that analysed sequences have no frequency component in common. Multiple cross-spectral functions of five different functional groups of proteins are represented in Figure 2.5. The regulatory DNA sequences were analysed in the same way.

The following fundamental conclusion was drawn from these studies: Each specific biological function of protein or regulatory DNA sequences is characterised by a single frequency.

These results are summarised in Table 2.2, where each functional group of proteins or DNA regulatory sequences is shown with its characteristic frequency and corresponding S/N within the multiple cross-spectral function.

It has previously been shown that a strong correlation exists between the amplitude spectrum of the numerical representation of macromolecular sequences and the corresponding biological function. Each RRM frequency characterises one biological function. The significance of a biological function is primarily recognition between participants in a particular biological process. It is assumed that RRM characteristic frequencies represent a relevant parameter of the recognition process

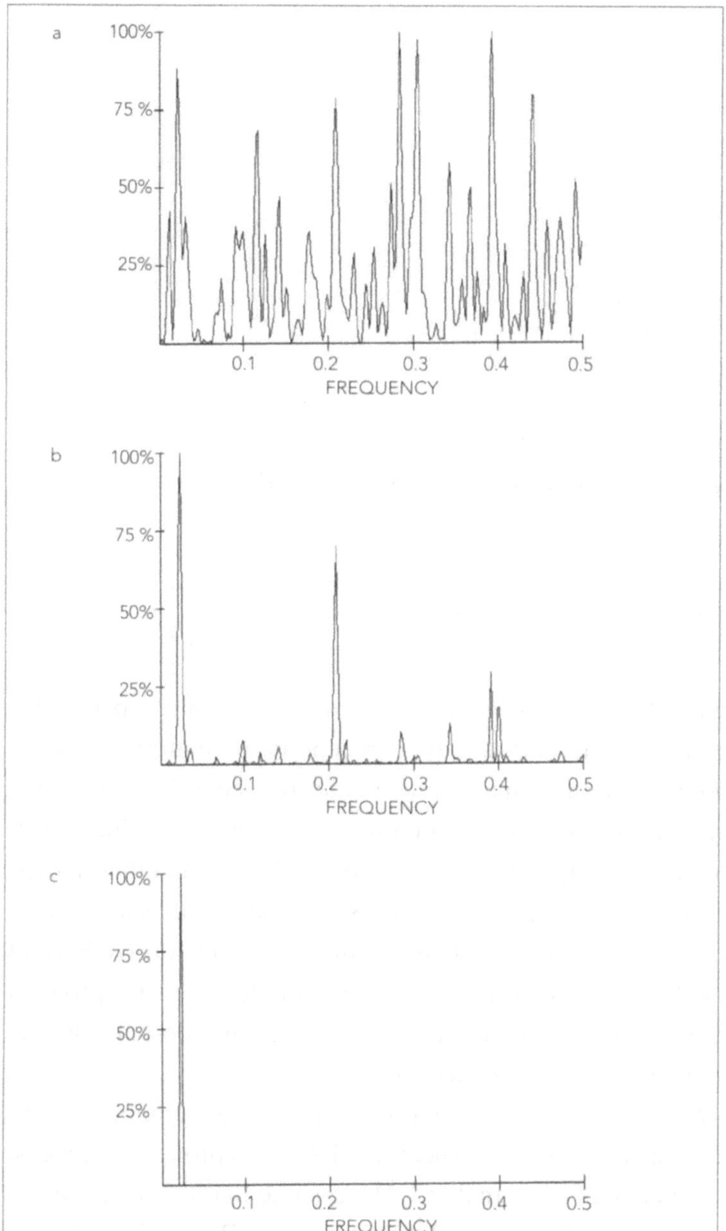

alpha human hemoglobin

human hemo-globins

hemoglobins

Figure 2.3. Evaluation of the consensus spectrum of one functional group of proteins: hemoglobins

(a) Single spectrum of human α-hemoglobin. (b) Multiple cross-spectral function of 4 human hemoglobins (α, β, δ and γ). (c) Multiple cross-spectral function (concensus spectrum) of all 187 hemoglobins.

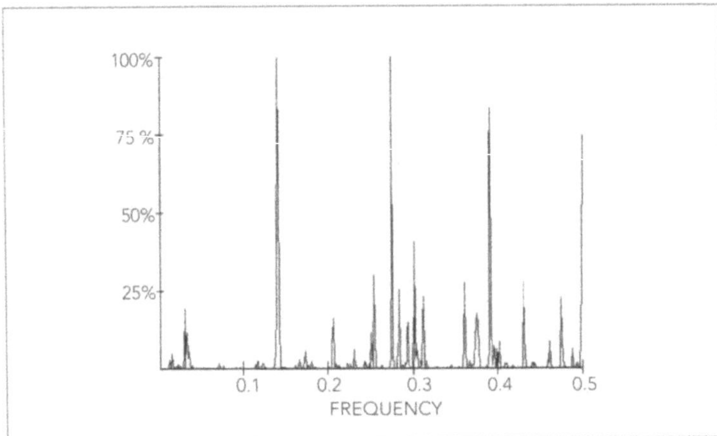

5 non-related proteins

Figure 2.4 Multiple cross-spectral function of five functionally non-related proteins
Human α hemoglobin, rat β fibroblast growth factor (FGF), human tumor necrosis factor (TNF), protein A and CD4 lymphocyte membrane protein. The resulting spectrum does not show a definite prominent peak.

between macromolecules. If the characteristic frequency is a resonant parameter for recognition between a protein and its target (receptors, regulatory DNA sequences etc.), both should have the same characteristic frequency. This has been shown in a number of examples of protein-protein interactions, including interleukins and their receptors [24, 29], nerve growth factors (NGFs) and their receptors [26, 46], and FGFs and their receptors [25, 34]. Some additional examples, including proteases and protease inhibitors and insulin-like growth factors (IGFs) and IGF-binding proteins, will be presented later. All these examples show that proteins and their interacting targets (receptors, binding proteins or inhibitors) display the same characteristic frequency in their interactions.

The situation was the same when DNA binding proteins were compared with their DNA targets. For comparison between spectra of DNA and protein sequences, the distance between nucleotides is normalised. We assume that the distance between amino acids in any protein is equal and that it is 3.8 Å. By similar assump-

Figure 2.5 Multiple cross-spectral function (consensus spectrum) of five different functional groups of proteins ▶
(a) Hemoglobins; (b) myoglobins; (c) lysozymes; (d) FGFs; and (e) cytochrome C. The multiple cross-spectral function of each group of proteins has a prominent peak representing the frequency characteristic to the biological function.

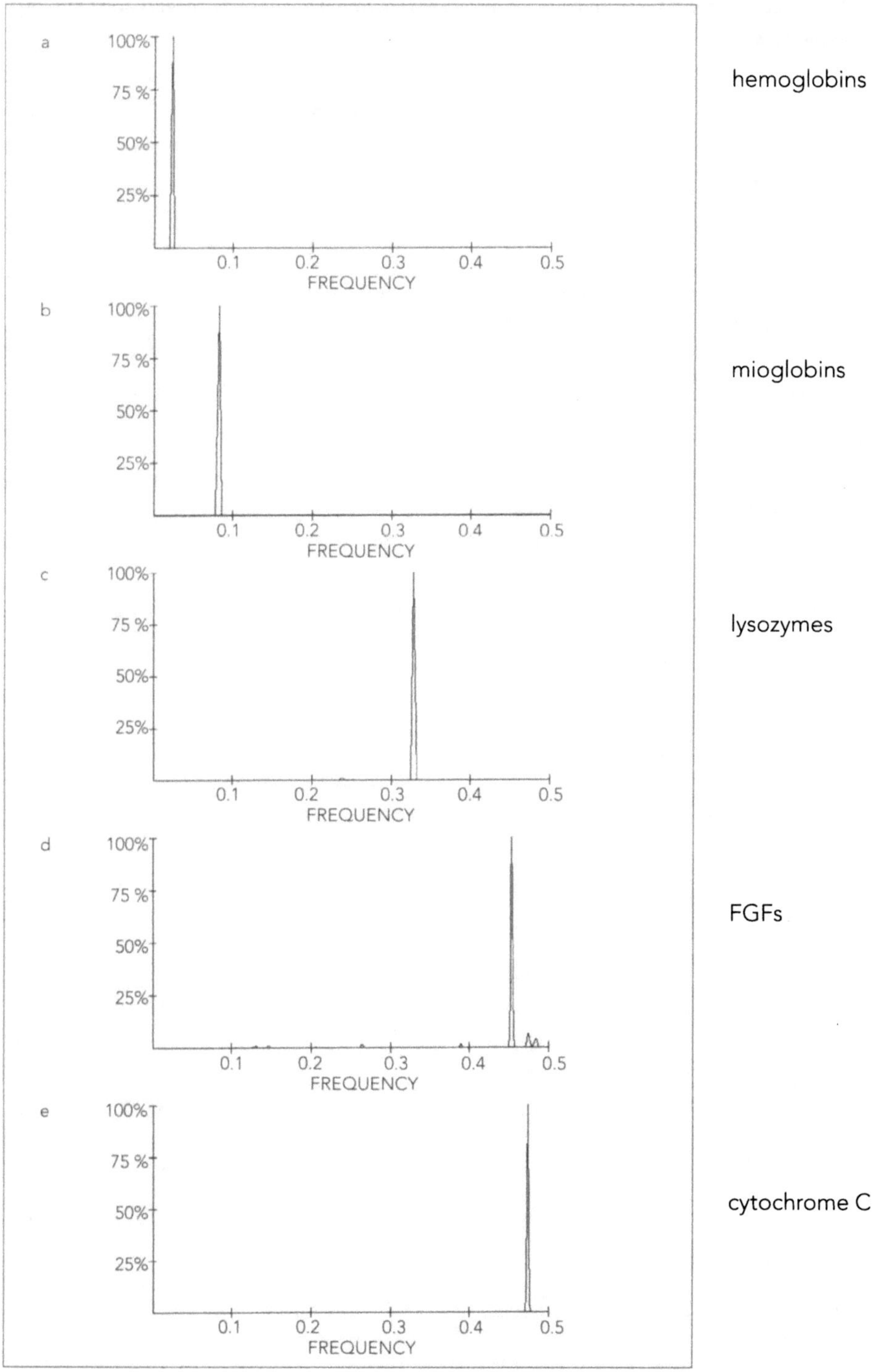
a
100%
75 %
50%
25%
0.1
0.2
0.3
0.4
0.5
FREQUENCY
hemoglobins
b
100%
75 %
50%
25%
0.1
0.2
0.3
0.4
0.5
FREQUENCY
mioglobins
c
100%
75 %
50%
25%
0.1
0.2
0.3
0.4
0.5
FREQUENCY
lysozymes
d
100%
75 %
50%
25%
0.1
0.2
0.3
0.4
0.5
FREQUENCY
FGFs
e
100%
75 %
50%
25%
0.1
0.2
0.3
0.4
0.5
FREQUENCY
cytochrome C

Table 2.2 Characteristic RRM frequencies for different functional groups of proteins and DNA regulatory sequences

Molecule type	Freq.	No. seq.	S/N	Error
DNA regulatory sequences				
promoters	.3437	53	128	.016
operators	.0781	8	44	.008
SOS operators	.4687	5	13	.050
enhancers	.0488	10	467	.024
Protein sequences				
ACH receptors	.4922	21	137	.002
actins	.4800	12	163	.002
amylases	.4121	12	170	.002
bacterial repressors	.0839	4	56	.004
cytochrome B	.0590	16	201	.004
cytochrome C	.4765	45	127	.004
fibrinogens	.4423	5	99	.001
glucagons	.3203	13	71	.034
growth factors	.2929	105	200	.016
ins.-like(IGF I,II)	.4922	12	72	.008
IGFBP (hum.)	.1602	6	172	.001
FGFs	.4512	7	121	.005
NGFs	.4040	8	192	.008
heat shock proteins	.0947	10	326	.005
hemoglobins	.0234	187	119	.008
homeo box proteins	.0459	9	100	.001
interferons	.0820	18	117	.008
kinases	.4297	8	71	.003
lysozymes	.3281	15	124	.004
myoglobins	.0820	49	128	.004
myosins	.3400	11	201	.004
neurotoxins	.0703	16	60	.004
oncogenes	.0313	46	468	.004
phage repressors	.1054	4	51	.005
phospholipases	.0430	29	115	.004
protease inhibitors	.3555	27	203	.008

proteases	.3770	80	511	.004
trypsins,	.3447	18	257	.004
chymotrypsin	.2363	5	35	.004
serine prot.	.4609	41	504	.004
repressors	.0990	25	198	.008
restriction enzymes	.2910	3	36	.004
signal proteins	.1406	5	31	.016

tion, the distance between nucleotides is accepted to be 3.4 Å. Therefore, the scale value for the distance between nucleotides is d(n) = 3.4/3.8 = 0.89. The maximal frequency in DNA spectra is F(d) = 1/2d(d) = 0.56, and the frequency resolution is f(d) = 1/Nd(d) f(d) = 1.14/N.

The consensus RRM spectrum of a group of 10 bacterial and phage promoters [13] is shown in Figure 2.6a.The scale value for the distance between nucleotides is taken into account as described above. The consensus RRM spectrum of a group of 10 bacterial and viral RNA polymerases is shown in the Figure 2.6b. Both groups of sequences are characterised by the same frequency, f = 0.335. The same observation was made with bacterial operators [13] and their corresponding repressors [24], as well as for enhancers and corresponding binding proteins (GKR) [27]. These results show that distinct characteristic frequencies exist for promoters and operators as DNA regulatory segments as well as for RNA polymerases and repressors as the respective binding proteins.

The model of recognition and activation between cell macromolecules is presented in Figure 2.7. It is assumed in this model that one protein can participate in more than one biological process; i.e., it may have more than one biological function. Each biological process is characterised by one frequency. A protein and its target have different biological roles, but they participate in the same biological process. This means that they should interact at the same frequency. From this point of view we can say that RRM frequencies characterise biological processes. The biological function (activity together with the mode of activity of the molecule) is characterised by frequency and by phase at this frequency. If the RRM characteristic frequency is the relevant parameter for recognition between proteins and their targets, then they can be expected to have the same characteristic frequency.

The question now arises : What is the difference between proteins of one biological function and their targets? Further research leads us to the conclusion that the answer to this question lies in the phase spectra of numerical representations of genetic sequences.

Proteins and their targets have the same characteristic frequency but ideally opposite phases (phase difference close to π) at this frequency. For example, in the case of the human interleukin-2 (IL-2) and its receptor, the phase difference was 2.77 rad [24], and in the case of NGFs and their receptors, the average phase difference was 3.96 rad [26, 46], both close enough to π. For the protein-DNA interaction, it was also observed that the phase difference at the characteristic frequen-

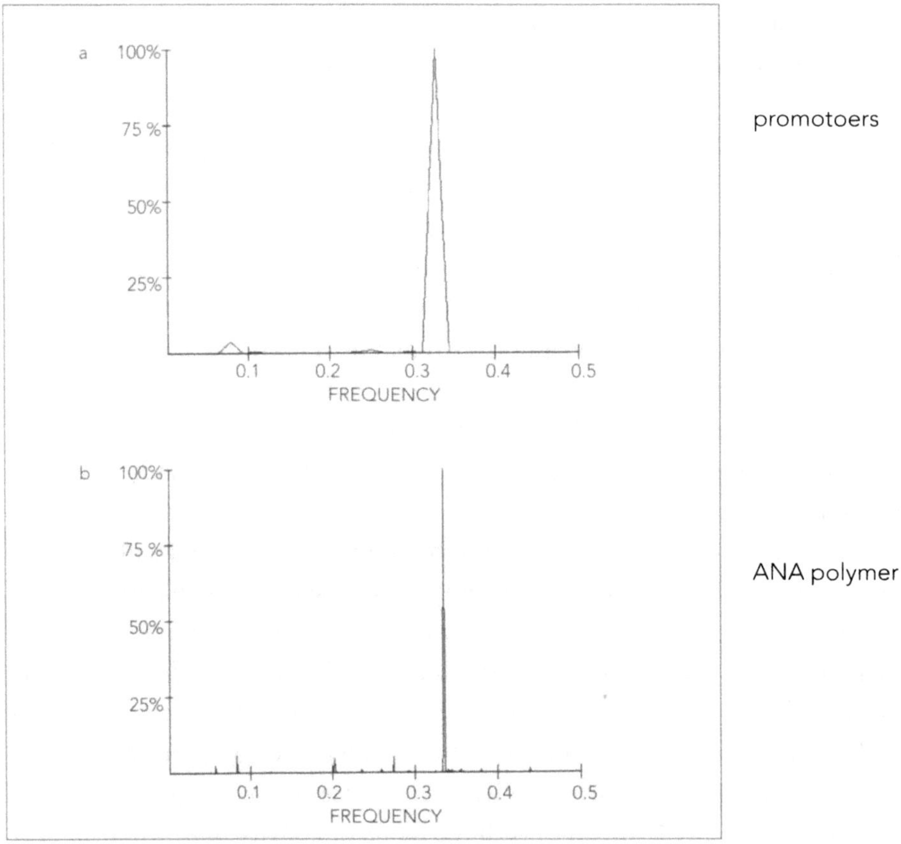

Figure 2.6 Comparison between the RRM frequency characteristics of interacting proteins
(a) The consensus RRM spectrum of a group of 10 bacterial and phage promoters; (b) the consensus RRM of a group of 10 bacterial and viral RNA polymerases. Both groups of sequences are characterised by the same frequency, f = 0.335, indicating that interaction between proteins is occuring at this frequency, i.e., interaction is a resonant process.

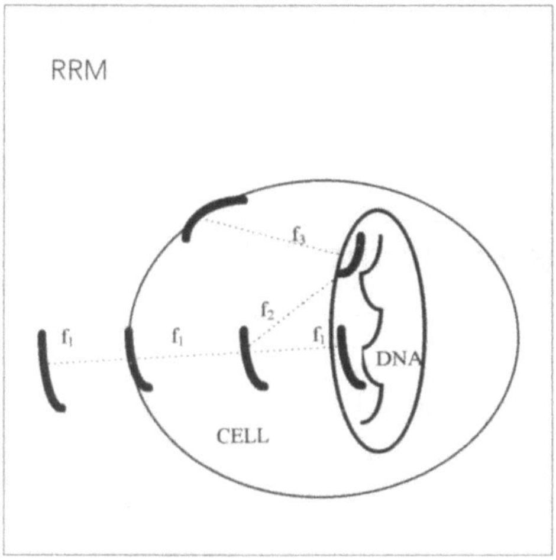

Figure 2.7 Schematic presentation of the RRM applied to a cell activation process
Macromolecules (proteins) P_i *recognise and interact with each other or with DNA regulatory segments at characteristic frequencies* f_i*. Each process has its characteristic frequency. However, the same frequency can characterise, within one biological function, a whole cascade of interactions, from extracellular through cytoplasmatic to nuclear and DNA interactions. In addition, one protein can be involved in two different functions or interactions, as a target or activator, and thus it can exhibit more than one characteristic frequency.*

cy for a particular interaction is close to π. For example, with four *Escherichia coli* promoters, phases at the characteristic frequency ranged from -2.88 to -3.31 rad with an average of -3.06 rad, while for the corresponding RNA polymerases, phases ranged from -0.52 to 1.83 rad, with an average phase difference of 1.95 rad. Similarly, for operators and their corresponding (binding) repressors, the phase difference was found to vary from 1.5 to 2.4 rad, again with an average of 1.95 rad. Although the phases are not quite opposite, the phase differences vary between 2 rad and -2 rad, which can be considered to be close to π. In addition, there are some indications that the phase differences point to the binding affinity and strength between interacting macromolecules. Thus, phase differences can identify species specificity of some interactions where there is no cross-reactivity between different species, although proteins within different species do have the same characteristic frequency. Thus, it is possible to conclude that the amplitude spectrum of the numerical representation of linear macromolecules contains information on the biological process/interaction in which molecules can be involved, while the phase spectrum describes the manner and the strength of this process or interaction.

According to the results presented, it can be seen that the RRM is built based on the following premises:

1. Linear macromolecules with the same biological function or interaction share a common characteristic frequency component.

2. Proteins and their target macromolecules have the same characteristic RRM frequency but opposite phases at that frequency.

2.2 Problems in FFT calculations within the RRM

Proteins are linear macromolecules built up of sequentially linked amino acids. The RRM analyses these linear sequences, transforming them into the frequency domain using the Fourier Transform (FT) calculated with the algorithm known as Fast Fourier Transform (FFT). Although FFT was generally used, in some cases we encountered problems using it. Here, we compare and discuss results obtained using FFT with different sequence lengths and DFT for the same groups of proteins.

For a more rapid calculation of the RRM characteristic frequency, FFT is used to obtain a single spectrum of each particular protein sequence [43]. These single spectra are then multiplied to obtain multiple cross-spectra for each protein functional group. In most cases there is no significant difference when DFT is used instead of FFT. For example, when comparing 90 hemoglobin sequences from different origins, the same characteristic frequency ($f = 0.024 \pm 0.008$) is obtained using FFT (Fig. 2.8a) and using DFT (Fig. 2.8b).

However, in some examples, there is a discrepancy between results obtained using DFT and results obtained using FFT [47]. For instance, in a comparison of 56 myoglobin sequences from different origins, the longest sequence is 154 amino acids. With FFT analysis, every sequence is extended with zeros up to the next higher number of terms equal to a power of 2, e.g. for myoglobins up to $2^8 = 256$ points. The multiple cross-spectrum obtained for the whole group of myoglobins reveals one prominent peak at the frequency $f_1 = 0.082 \pm 0.008$ (Fig. 2.9a).

But when the same calculations are carried out with a sequence length of $2^9 = 512$, a completely different common frequency, $f_2 = 0.252 \pm 0.008$, appears in the multiple cross-spectrum (Fig. 2.9b). If we observe the single spectrum of each myoglobin sequence, we can see that each does contain both frequencies. However, when the sequence is extended with more zeros, the second frequency becomes

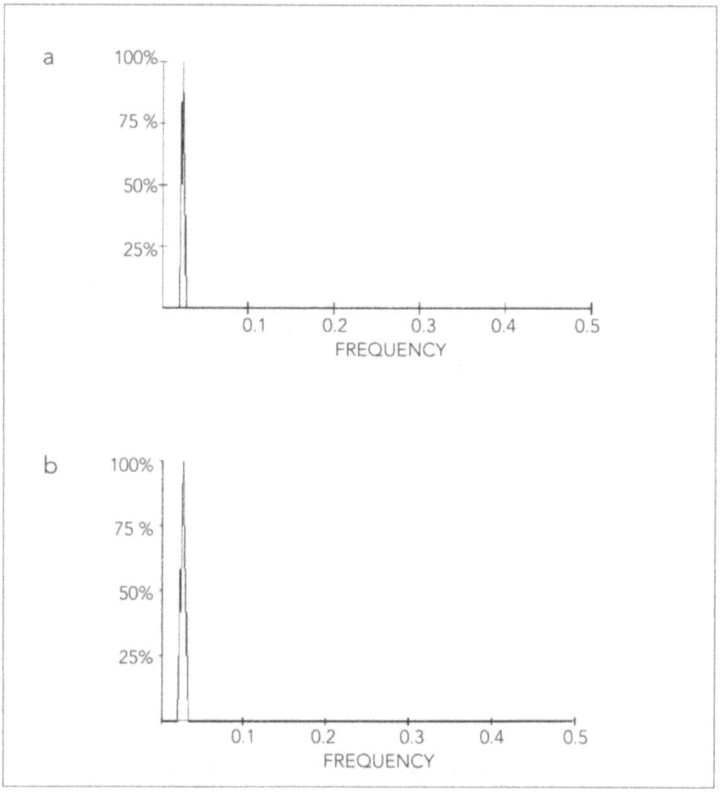

Figure 2.8

(a) Multiple cross-spectral function of 90 hemoglobin α and β sequences from different origins, obtained using the FFT algorithm. The prominent frequency peak is at f = 0.024 ± 0.008 with S/N = 127. (b) Multiple cross-spectral function of 90 hemoglobin α and β sequences from different origins, obtained using the DFT algorithm. The prominent frequency peak is at f = 0.024 ± 0.008 with S/N = 77.

slightly more prominent. Because this appears in each sequence, the difference is drastically enhanced using the multiple cross-spectrum.

DFT analysis was also applied to myoglobins to calculate initial single spectra. When these spectra are multiplied, the frequency f_2 is found to be prominent in the resulting multiple spectra (Fig. 2.9c).

These results show that when different techniques of calculating discrete spectra are applied to some protein groups, minor differences in the single spectra can cause significant differences in the multiple spectrum. Consequently, significant errors in the calculation of the characteristic RRM frequency can occur.

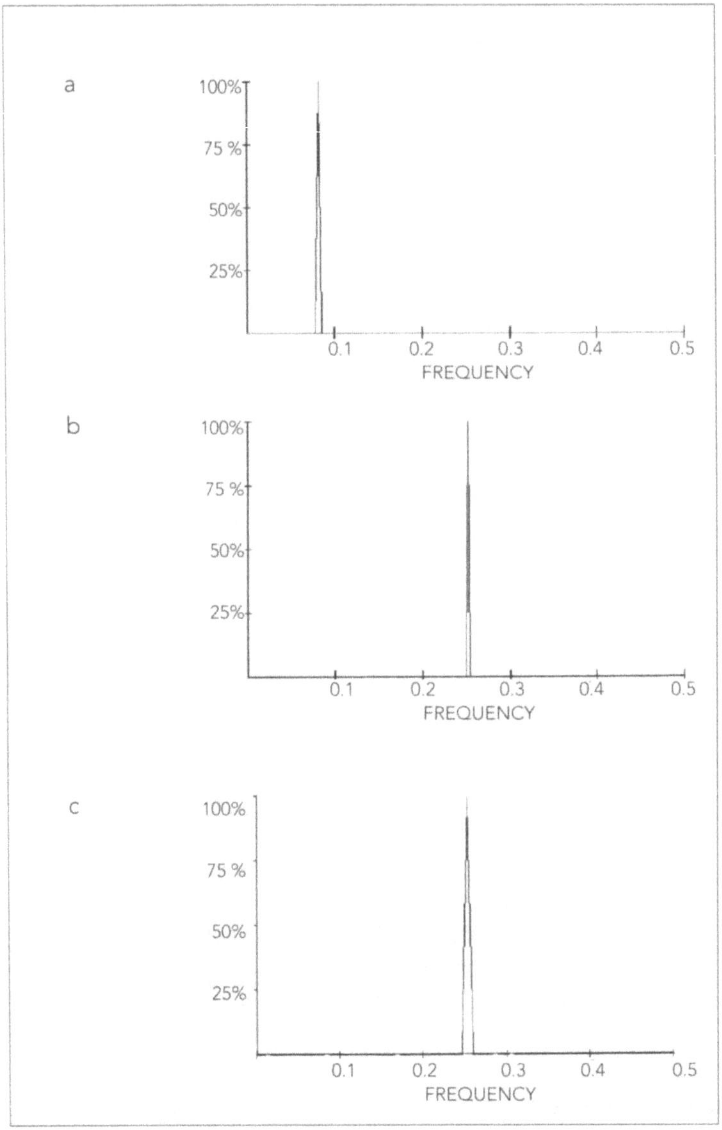

Figure 2.9
(a) Multiple cross-spectral function of 56 myoglobin sequences from different origins, obtained using FFT on a sequence length of 2^8 = 256. The prominent frequency peak is at f = 0.082 ± 0.008 with S/N = 128. (b) Multiple cross-spectral function of 56 myoglobin sequences from different origins, obtained using FFT on a sequence length of 2^9 = 512. The prominent frequency peak is at f = 0.252 ± 0.008 with S/N = 255. (c) Multiple cross-spectral function of 56 myoglobin sequences from different origins, obtained using DFT with zeros added to a full sequence length. The prominent frequency peak is at f = 0.253 ± 0.008 with S/N = 78.

No signal can be both time-limited and band-limited; i.e., the spectrum of a band-limited signal should be infinitely long. Our signal is already discrete and of limited length because proteins are of limited length, i.e. with a limited number of sequentially linked amino acids. On the other hand, the limited length of the signal generates an abrupt termination to this data series. In effect, this truncation is a multiplication of the signal series with a rectangular data window of length N. Hence, in the frequency domain, the spectrum is actually the convolution of the real spectrum of EIIP and the spectrum of that data window which is a sinc function. Then, a leakage effect occurs, which both blunts the sharpness of the spectrum and limits its spectral resolution [43].

FFT is a fast computing algorithm for DFT. The difference between FFT and DFT is that FFT needs signals with an exact integer power of 2. Consequently, in many cases, zeros are appended to the data. With EIIP, padding with different numbers of zeros is needed for different protein sequences in order to make the whole group the same length. Signals with identical length within the protein group are needed for further multiple cross-spectral analysis. Zero padding should be used on the premise that the signal is indeed zero outside the record length. Zero padding in the time-space domain in no way improves the resolution of the spectra; it simply provides a method of interpolation in the frequency domain [43]. As a result, due to the combination of aliasing, leakage and zero padding, only estimated spectra results could be obtained.

To eliminate the difference between multiple cross-spectra derived with the direct FFT algorithm with different powers of 2, an improved, indirect, Blackman-Tukey method was used to calculate spectra [48]. The Blackman-Tukey method involves two steps: calculation of the autocorrelation function of the original sequence and then a Fourier transformation of this autocorrelation function, which gives the energy density spectrum of the original data [49]. In addition, data windowing was introduced [43] to smooth the edges at the end of the sequence, and to decrease problems arising from zero padding.

The spectra obtained using the Blackman-Tukey method combined with the Hanning window are more alike in shape and have a common peak at frequency 0.082, though different powers of 2 are used for the FFT. In further analysis, multiple cross-spectra were obtained for four groups (myoglobin, hemoglobin, cytochrome C and glucagon) of functionally related proteins using both the direct method and the Blackman-Tukey method. In most cases, there was no significant difference among the results obtained with different powers of 2 used in radix-2

FFT and those simply derived using DFT algorithms via the direct method. However, as described above, for some groups of proteins like myoglobin [47], it has been found that there is a significant difference in the results obtained through these direct methods.

With the results of the Blackman-Tukey method combined with windowing, and since the single spectrum is now independent of the power of 2, the final multiple cross-spectral function defined by RRM should also be independent of the power of 2. This is presented here in an example of myoglobins. The multiple cross-spectra of 60 myoglobins revealed a common characteristic frequency at $f_1 = 0.082$, and this consistent result could be considered as the more accurate one [48].

3 The Physical Meaning of the RRM

Mathematically obtained RRM characteristic frequencies of macromolecular biological functions can be considered as resonant energy transfer between interacting molecules. This energy can be transferred through oscillations of a physical field which may be electromagnetic in nature.

The correlation between the amplitude spectrum of the numerical representation of linear macromolecules (proteins and DNA) and the corresponding biological function presented previously can lead to a completely new approach to the study of protein dynamics. Each frequency in the RRM characterises one biological function. To grasp the meaning of characteristic frequency, it is important first to understand what is meant by the biological function of proteins. Each biological process involves a number of interactions between proteins and their targets (other proteins, DNA regulatory segments or small molecules). Each of these processes involves energy transfer between interacting molecules. These interactions are highly selective, and the selectivity is encoded within the protein structure. Proteins and their protein or DNA targets have been analysed to find out whether RRM characteristic frequencies denote a parameter which describes this selectivity between interacting molecules. It has been shown that proteins and their DNA or protein targets share the same characteristic frequencies [24, 25, 27, 29, 46] but with opposite phases [24, 25, 46] for each pair of interacting macromolecules. Thus, it can be postulated that RRM characteristic frequencies characterise not only general functions but also recognition between a particular protein and its target (receptor, ligand etc.). Because this recognition arises from the matching of periodicities within the distribution of energies of free electrons along the interacting proteins, it can be regarded as resonant recognition.

 ©1997, Birkhäuser Verlag Basel

3.1 Charge transfer through the protein backbone

There is evidence that proteins have certain conducting or semiconducting properties. Thus, charge moving through the protein backbone and passing through different energy stages caused by different amino acid side groups can produce sufficient conditions for specific electromagnetic radiation or absorption. The frequency range of the related field depends on charge velocity estimated to be 7.87×10^5 m/s and on the distance between amino acids in protein, which is 3.8 Å. The frequency range obtained for protein interactions is 10^{13}–10^{15} Hz. This estimated range includes infrared, visible and ultraviolet light and is consistent with the linear correlation (a) between absorption characteristics of light-absorbing proteins and their characteristic RRM frequencies and (b) between frequency-selective light effects on cell growth and characteristic RRM frequencies of growth factors.

The RRM assumes that characteristic frequencies are responsible for resonant recognition between macromolecules at a distance. Thus, these frequencies have to represent oscillations of some physical field which can propagate through water dipoles. One possibility is that this field is electromagnetic in nature. The whole process of protein interactions can be observed as the interaction between the transmitting and receiving antennae of a radio system. There is evidence that proteins and DNA have certain conducting properties [16-20]. If so, charges would be moving through the backbone of the macromolecule and passing different energy stages caused by different side groups of various amino acids or nucleotides. This process provides sufficient conditions for the emission of electromagnetic waves. Their range of frequency depends on charge velocity, which in turn depends on the nature of charge movement (superconductive, conductive, soliton transfer etc.) and on the energy of the field that causes charge transfer. The nature of this physical process is still unknown, though some models of charge transfer through the backbone of macromolecules have been accepted [18, 19]. Each of these shows that charge transfer through the backbone of macromolecules is possible, but the complexity of the system does not allow precise calculations of charge velocity. Furthermore, it is not possible to find any correlation between charge transfer and the biological function of macromolecules. The significance of the whole process is proposed below, although the author is well aware that conductive electron transfer caused by the difference of the free electron potentials (EIIP) at the N- and C-terminals of the protein is beset by conditions that are not well understood. According to the pseudopotential, this potential energy difference is

$$W = W(COOH) - W(NH_2) = 0.13 \text{ Ry}$$

This energy difference allows for a maximum velocity of the electrons, which is equal to

$$V_{max} = \sqrt{(2eW/m)}$$

where e is the electron charge, and m is electron mass. Therefore,

$$V < 7.87 \times 10^5 \text{ meters/second}$$

An inherent assumption is that amino acids in the protein are equidistant and the distance is

$$d = 3.8 \text{ Å}$$

Thus, the maximum frequency that could be emitted during electron transfer is

$$F_{max} < V/(2d)$$
$$F_{max} < 1 \times 10^{15} \text{ Hz}$$

while the corresponding wavelength is

$$L_{min} > 330 \text{ nm}$$

The minimum frequency that could be emitted depends on the total length of the protein

$$F_{min} = 2F_{max}/N$$

where N is the total number of amino acids in the protein. For example, with proteins 200 amino acids in length, the minimum frequency is

$$F_{min} < 1 \times 10^{13}\ \text{Hz}$$

and the corresponding wavelength is

$$L_{max} < 30{,}000\ \text{nm}$$

The range from 30,000 nm to 300 nm is very wide, starting from the very low infrared through the visible to the ultraviolet regions. This also covers the entire range of sunlight irradiation on the earth's surface [50], and it corresponds to the radiation necessary for the creation of life. It is also important to notice that the upper limit of the proposed frequency range is below the frequency range of the resonant absorption peak of single amino acids. This range is only an estimate of the possible frequency range of the field that could be emitted by charge transfer through the backbone of proteins. A more precise estimation would require different biological experiments. Known experimental results were needed to reach this aim and are explained here.

3.2 Correlation of the RRM characteristic frequency with absorption spectra of light-absorbing proteins

The estimated electromagnetic energy levels were initially investigated by comparison of the absorption spectra of some groups of chromophore-bearing proteins with their corresponding characteristic RRM frequencies. A linear correlation was obtained with a regression coefficient of K = 201.

The RRM is based on the concept that biological function represents a resonant energy transfer from one biomacromolecule to another [24, 25, 51]. However, the function of some proteins is directly related to the absorption of visible light of defined wavelengths. Thus, a correlation is expected between the absorption spectra of proteins and their RRM spectra with a predetermined frequency range. Such a correlation has already been demonstrated in previous studies [24, 25, 30], and it is summarised here.

Light is absorbed effectively on a prosthetic group bound to a protein, but frequency selectivity in this process is defined by the protein itself. For example, rhodopsins, which have the same prosthetic group within proteins of similar but different primary structures, can absorb different wavelengths (there are three different variants of rhodopsins, one for the absorption of each of the colours red, blue and green [52]). Since the biological function of these molecules is to absorb particular wavelengths (colours), their grouping was achieved on this basis. Thus the rhodopsin responsible for the absorption of blue light was compared with the bioluminescent protein aequorin, which absorbs at similar wavelengths [53]. In the corresponding multiple cross-spectrum there is only one prominent peak at frequency 0.475 ± 0.004, and this is the peak most likely to be related to the absorption of blue light. In the same way, it is estimated that the frequency of 0.355 ± 0.004 is related to the absorption of green light, and the frequency of 0.346 ± 0.004 to the absorption of red light. The numerical frequencies obtained similarly by the RRM for various other groups of visible light-absorbing proteins are compared with their corresponding characteristic absorption frequencies in Table 3.1.

Table 3.1 Correlation between characteristic RRM frequencies and visible infrared absorption maxima of different groups of photosensitive proteins

Protein group	Wavelength (nm)	Frequency (cm^{-1})	Relative RRM frequency	K
Cyt C	415 ± 20	24,096 ± 117	0.473 ± 0.003	196
Blue	430 ± 25	23,256 ± 1278	0.475 ± 0.004	204
Green	540 ± 30	18,518 ± 974	0.355 ± 0.004	192
Red	570 ± 30	17,544 ± 877	0.346 ± 0.004	197
Purple	860 ± 35	11,628 ± 455	0.281 ± 0.02	241
Flavodoxin	470 ± 30	21,275 ± 1275	0.379 ± 0.004	178

The groups are formed according to their characteristic absorption maxima: Cyt C (cytochrome C); blue (group of proteins absorbing blue light: blue rhodopsin and bioluminescent proteins); green (green light-absorbing chlorophylls and rhodopsin); red (red light-absorbing rhodopsin); purple (purple light-absorbing proteins from purple bacteria). The second column represents the wavelengths (nm) of characteristic absorption of the groups of photosensitive proteins, while the third represents the same data in cm^{-1} units. The fourth column contains characteristic RRM frequencies f of the same groups of proteins, while the fifth column contains the coefficient K, the scaling factor between the numerical frequency space of RRM and the wavelengths λ (nm) of the real frequency space: $\lambda = K/f$. The mean value of K is 201 with a standard deviation of 15%.

A result of considerable significance is that the scaling factor between these two sets of data is almost constant at the mean value of K = 201. Thus, a strong linear correlation would seem to exist between the numerical characteristic frequencies defined by the RRM and the experimentally determined frequencies corresponding to the absorption of electromagnetic radiation of such proteins. From this correlation it can be observed that the full range of wavelengths which can be related to RRM characteristic frequencies is over 400 nm. This finding is in complete accord with the frequency range previously associated with the RRM spectra and calculated from the charge velocities through the protein backbone. It can now be inferred from both correlations that approximate wavelengths in real frequency space can be calculated from the RRM characteristic frequencies for each biologically related group of sequences. Furthermore, these calculations can be used to predict the wavelength of visible and near-infrared irradiation which may have a biological effect.

The situation is slightly different for cytochrome B, whose characteristic RRM frequency was found to be at 0.059, and for myoglobin, whose characteristic frequency is 0.082. According to the RRM, it is expected that these proteins would have absorption at 3400 nm and 2451 nm, respectively. On the other hand, it is known that there is an absorption in the visible light range for cytochrome B at 423-427 nm and for myoglobin at 533-582 nm. As there is not any data on absorption in cytochrome B and myoglobin within the expected very low frequency range (order of value of thousands of nanometers), we cannot conclude whether these proteins are an exception or whether they possibly have an even stronger absorption at the predicted low frequencies.

3.3 Correlation of RRM characteristic frequency with low-intensity light effects on cell proliferation

A comparison of optimal wavelengths for the effect of low-intensity light irradiation on cell proliferation and RRM frequency characteristics of peptide growth factors involved in cell proliferation revealed a linear correlation. Strikingly, this correlation exhibited the same coefficient (scaling factor) K = 201 ± 14 between the RRM frequency space and corresponding light wavelengths in nanometers.

The frequency selectivity of many light-induced biological processes is caused directly by the quantum-mechanical energetic states of photosensitive molecules. There is considerable evidence to suggest that the induced change of the energy states of biomacromolecules (such as the effect of ultraviolet and visible light on primary photoacceptors) leads to the modulation of some biological processes in cells [54]. There is also considerable evidence that low- intensity light irradiation at precisely defined wavelengths (frequencies) can produce defined, frequency-dependent effects on living systems *in vivo* or *in vitro*. The correlation between the frequency selectivity of light-induced biological processes and RRM characteristics of biomacromolecules involved in these processes has been described for the case of cell growth and proliferation [30]. These results are summarised in Table 3.2, in which light irradiation frequencies which have been shown to produce ef-

Table 3.2 Correlation between characteristic RRM frequencies of five groups of growth-regulating factors and the low-intensity light irradiation frequencies which produce significant effects on cell growth

Growth factor group	Effect observed	RRM frequency, f_c	Expected wavelength, λ_c (nm)	maximum effect (nm)
IGFs	DNA synthesis	0.492 ± 0.008	406 ± 6.5	400
FGFs	DNA synthesis	0.453 ± 0.004	441.5	441.6
	Therapeutic effects			
Insulins	DNA synthesis	0.344 ± 0.16	581.4 ± 10	552
Growth factors	DNA synthesis	0.293 ± 0.016	682.6 ± 35.4	633
(EGF, CSF, GH,	ATP synthesis			650
PLF)	Therapeutic effect			633

The second column lists biological effects observed under laser irradiation of a particular frequency shown in the fifth column. The third column contains RRM characteristic frequencies of the different growth factor groups listed in the first column. The fourth column list the real frequency space wavelengths according to the relation $\lambda = 201/f_{RRM}$, where f_{RRM} is the calculated numerical RRM frequency. It can be observed that expected wavelengths calculated with the RRM are in complete agreement with the wavelengths which can produce maximum biological effects.

fects on cell growth are compared with characteristic RRM frequencies for different groups of growth factors normally involved in these cell proliferation processes. Linear correlation was again obtained with the same regression factor as that found for light-absorbing molecules. The most illustrative comparison is the one between expected wavelengths, calculated from RRM frequencies using correlations obtained with light-absorbing molecules, and measured wavelengths which can produce certain biological effects.

The graph of Figure 3.1 shows the results of the comparisons between RRM characteristic frequencies for light-absorbing proteins and their absorbing characteristics and between RRM frequencies of different groups of growth factors and frequencies of light irradiation which can produce effects on cell growth. Both cases show the same linear correlation. Strikingly, this correlation already exhibited the same coefficient (scaling factor) of K = 201 between the RRM frequency space and corresponding light wavelengths in nanometers [24]. This can be represented as follows:

$$\lambda = K/f_{RRM}$$

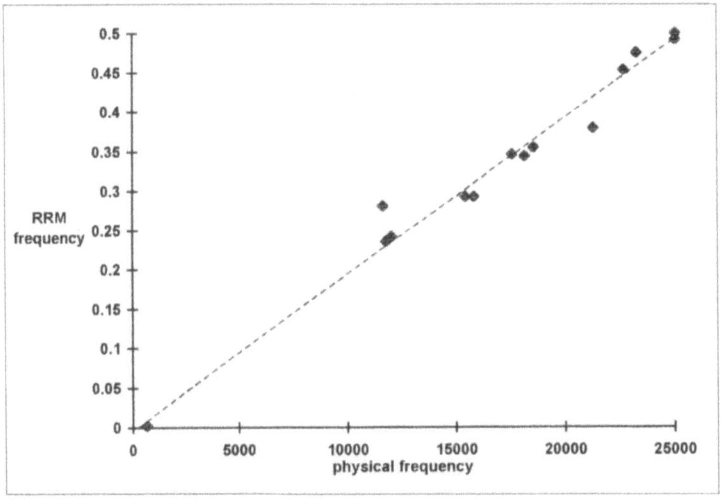

Figure 3.1 Correlation between the real, physical frequency space and RRM frequency space

Each point in the graph represents one group of proteins, either light-absorbing proteins or growth factors, with their RRM characteristic frequency on the x-axis and their corresponding physical frequency characteristics on the y-axis.

where λ is the wavelength in nanometers of the light irradiation which can influence a particular biological process (cell growth or light absorption), f_{RRM} is the numerical frequency obtained by the RRM and K is the estimated coefficient of the linear correlation obtained.

3.4 Correlation between predicted and measured characteristic frequency of chymotrypsin activation

Frequency-selective effects of light on biological processes of protein activation imply that protein activation involves energies of the same order and nature as electromagnetic irradiation of light. Consequently, this suggests that biological processes could be influenced by an external radiation of defined frequency. These phenomena are discussed here in terms of the RRM which proposes that protein activities (ie. protein interactions) are based on resonant electromagnetic energy transfer within a range of infra-red and visible light.

As discussed above, there is evidence that biological processes can be induced or modulated by irradiation with characteristic frequencies. This is directly caused by light-induced changes of the energy states of molecules and in particular of proteins. Consequently, one may infer that biological processes can be influenced by an external radiation of defined frequency. The strongest support for this idea emanates from an experiment described by Biscar [55] which shows that protease activity of α-chymotrypsin is significantly increased in the presence of a near infrared beam of defined wavelength [25].

The RRM was applied to the example of chymotrypsins, which belong to the group of proteases [56, 25]. Their common biological function is hydrolysis of proteins. Chymotrypsin is synthesised as one chain, chymotrypsinogen, which after cleavage results in a fully active enzyme, α-chymotrypsin. Cross-spectral analysis of five active chymotrypsins from different origins (bovine, human, rat, vop and oriental horne) revealed a common frequency component at $f = 0.236 \pm 0.004$ with $S/N = 35$. This frequency component can be considered as a characteristic specifically associated with chymotrypsin activation rather than with its protease activity. In comparison with other proteases, a different frequency ($f = 0.377 \pm 0.004$) was obtained as a general protease characteristic [24, 25, 28].

Using the relation between RRM frequency and light irradiation wavelength, $\lambda = 201/f_{RRM}$, obtained previously, the expected light wavelength which can influence chymotrypsin activation was calculated to be 851 nm. In addition, Biscar [55] has reported an experimental increase in chymotrypsin activity by a factor greater than 2 under infrared irradiation of a wavelength of 855 nm. The activity of the enzyme was unaffected (equal to the control in the absence of irradiation) outside the range 850-860 nm. As can be observed from our results, the light wavelength predicted by the RRM as characteristic of chymotrypsin activation is exactly within the measured range of chymotrypsin activation. These and all previous results showing correlation between RRM and real physical frequency space are summarised in Table 3.3.

Table 3.3 Light-absorbing proteins, growth factors and chymotrypsins with their absorbing/activating frequencies and calculated RRM frequencies

Computed (numerical) values are obtained computationally from the proposed electron velocities through the protein backbone. The coefficient of transformation between these two frequency spaces is almost constant and is about 200 (201 ± 14).

Protein group	Function	Physical freq. (nm)	Physical freq. (cm^{-1})	Calculated RRM freq.	K = freq. (nm) × RRM freq.
cyt. c	light abs.	415	24096.38	0.473	196
blue	light abs.	430	23255.81	0.475	204
green	light abs.	540	18518.51	0.355	191
red	light abs.	570	17543.85	0.346	197
hem.	light abs.	14770	677.0480	0.023	295
purple	light abs.	860	11627.90	0.281	241
flavodoxin	light abs.	470	21276.59	0.379	178
IGF	growth	400	25000	0.492	196
FGF	growth	441.6	22644.92	0.453	200
insulin	growth	552	18115.94	0.344	189
growth f. 1	growth	633	15797.78	0.293	185
growth f. 2	growth	650	15384.61	0.293	190
computed	numerical	400	25000	0.5	200
PDGF	growth	830	12048.19	0.242	200
chymotr.	protease	851	11750.88	0.236	200

These results strongly support the idea that RRM characteristic frequencies represent specific electromagnetic field oscillations within the infrared and visible range which are crucial for protein activity and interactions. These results also reinforce linear correlation between the RRM frequency space and wavelengths of biologically effective light irradiation with a correlation coefficient of K = 201. With this coefficient in mind, it is now possible to calculate wavelengths of light irradiation which, it is proposed, will affect other biological processes. The triggering or control of some biological processes in the cell via irradiation with light of a defined wavelength would also be possible and, indeed, does seem to occur.

3.5 Photoreceptors with similar structures but different absorptions can be distinguished using the RRM

Light-specific photoreceptor activations in plants are discussed here in terms of the RRM model, which proposes that proteins are activated with characteristic resonant frequencies. The RRM characteristic frequencies of red/far-red HY4 mutants were compared with the characteristic frequency of blue-light photoreceptor photolyases. A clear distinction between red/far-red and blue-light receptors was detected, in addition to their similar structures. Furthermore, complete agreement was achieved with the previously established correlation between RRM frequencies and protein-activating spectra.

In our previous work we postulated that bioactivity of proteins is based on protein excitation with resonant electromagnetic energies of characteristic frequency for each particular biological function [24, 25]. These frequencies are found to be within the range of very far infrared, through visible up to UV light. The postulates which are the basis of the RRM model have been tested on a number of examples, including light-absorbing proteins [24, 25], growth factor activation [25, 30] and enzyme activation [25, 56]. Here, we will use this theory to analyse light-specific responses in plants, in particular, HY4 mutants of *Arabidopsis thaliana* and a group of photolyases. Both groups of proteins have significant structural similarities [57], but the RRM approach made it possible to distinguish their different activation characteristics.

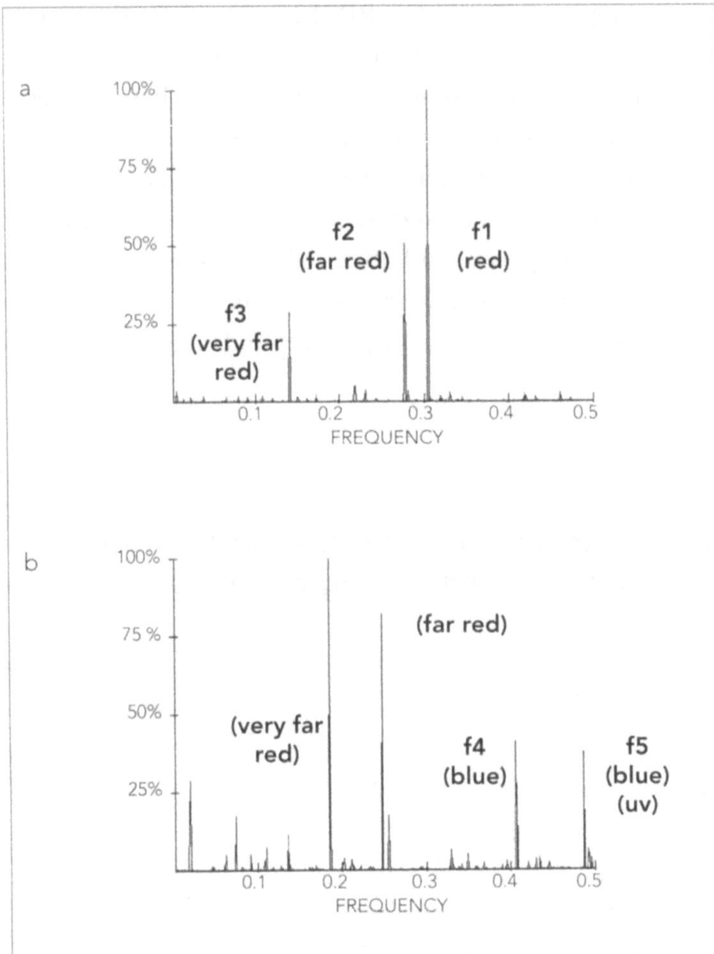

Figure 3.2
(a) Multiple cross-spectral function of four HY4 mutants. The following RRM frequency peaks occur: f_1 *= 0.3047,* f_2 *= 0.2773 and* f_3 *= 0.1460, with corresponding expected λ (nm)(λ = 201/*f_{RMM}*) 656 (red), 721 (far-red), 1422 (very far red), respectively. There are no peaks in the proposed green or blue range. (b) Multiple cross-spectral function of seven photolyases. The following RRM frequency peaks occur:* f_4 *= 0.4053 and* f_5 *= 0.4863, with corresponding expected λ (nm) (λ = 201/frrm) 411 (blue) and 493 (blue and UV), respectively. The other frequency peaks are in the far-red range. There are peaks in the proposed blue range.*

HY4 mutants are arabidopsis mutants of phytochrome photoreceptors. They are selectively insensitive to blue light during blue light-dependent inhibition hypocotyl elongation response, when they are still capable of responding to far-red and red light [57, 58]. When the RRM was applied to four different HY4 mutants (hy4, hy401, hy403 and hy404 [57]) it revealed three common characteristic frequencies: $f_1 = 0.3047$, $f_2 = 0.2773$ and $f_3 = 0.146$ (Fig. 3.2a). Using the relation between the RRM frequency and the light irradiation wavelength $\lambda = 201/f_{RRM}$ previously obtained, the light wavelengths which can be expected to characterise HY4 activation are $\lambda_1 \sim 656$ nm (red), $\lambda_2 \sim 721$ nm (far red) and $\lambda_3 \sim 1422$ nm (extremely far red), respectively. No peaks were found in the proposed blue or green regions of the spectrum. This clearly indicates that HY4 mutants, in line with the RRM postulates, are sensitive to activation only with red and far-red light but not with blue or green light, which is also in complete agreement with the experimental results [57, 58]. On the other hand, multiple cross-spectral function of seven different photolyases revealed four prominent peaks, two of them in the far- red and very far red regions and the other two in the blue region: $f_4 = 0.4053$ and $f_5 = 0.4863$ with corresponding wavelengths of $\lambda_4 \sim 411$ nm and $\lambda_5 \sim 493$ nm (Fig. 3.2b). Photolyases are characterised by two different absorption peaks: long waves at about 435 nm and short waves at about 380 nm [57]. Although two peaks in the multiple cross-spectra do not exactly denote short- and long wave wavelengths, the error is about 10%, which can be tolerated keeping in mind that both measurements and the RRM involve a number of approximations. Nevertheless, these results clearly predict two distinct absorption frequencies in photolyases, both within blue spectra [59].

3.6 Frequency range of macromolecular interactions

As is possible to see from the above results, the RRM can distinguish between structurally similar proteins which have different activation energies and frequencies. In addition, these results support the idea that the RRM characteristic frequencies represent specific energies within the infrared and visible range which are crucial for protein activation. This also confirms the linear correlation between the RRM frequency space and wavelengths of biologically effective light irradiation with a correlation coefficient K = 201. With this coefficient in mind, it is now possible to calculate wavelengths of light irradiation which can be proposed to influence other protein activation and biological processes. The relation between RRM and real frequency spaces is represented graphically in Figure 3.3, where RRM units are compared with nm and cm^{-1} for easier calculations between spaces.

Previous results could lead to the conclusion that each macromolecular biological process inside the living cell is characterised by one frequency within a very wide frequency range from the extremely low infrared up to the ultraviolet. Does this mean that basic intermolecular life processes are carried out at their own frequencies within this defined frequency range? Once, it was claimed that the reason for the transformation of inorganic to organic substances was solar heat and light [60]. It is known that heat from the sun and sunlight are electromagnetic radiation emitted in the frequency range from low infrared to ultraviolet. This is the same frequency range which is assumed here to be responsible for protein-protein and protein-DNA interactions, the main intermolecular processes of life.

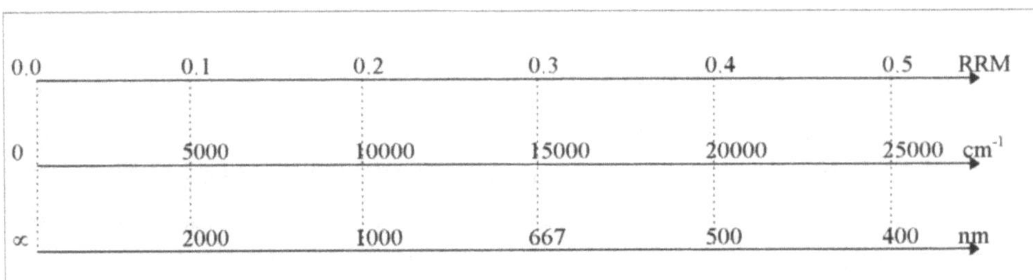

Figure 3.3 The relation between the RRM and physical frequencies in cm^{-1} and nm

Process	Frequency
heart activity	0.5 Hz
brain activity	1-30 Hz
muscular activity	100 Hz
thermal effects	10-100 MHz
DNA absorption	1-10 GHz
cell membrane	100-1000 GHz
intermolecular activity	10^3-10^6 GHz
absorption of amino acids and nucleotides	10^6 GHz
biological damage	$>10^6$ GHz

Table 3.4 Frequency ranges of different biological activities

The upper limit of the proposed range is just below the resonant frequency of amino acids themselves [61]. The lower limit of this range is above the resonant frequency of living elements which are larger than proteins (i.e. cell membrane [62] or the entire DNA [63]). The frequency range of intermolecular biological processes is presented in the following table together with frequencies characteristic of other biological processes.

The idea that every intermolecular biological process is carried out at its own defined frequency opens up the possibility of influencing biological processes at any level from bacteria to higher organisms. This could also lead to the discovery of new technologies that could be based on the principles of living systems.

4 Relation Between the RRM and 3D Protein Structures

It is known that proteins cannot express their biological function until they achieve a certain active 3D structure. In this chapter, the relation between RRM characteristic frequencies arising solely from the amino acid sequence and the active 3D protein structure is discussed. A number of examples are presented.

4.1 Definition of "hot spots" in terms of the RRM

Since the characteristic frequency f(x) correlates with the biological function x, the positions of the amino acids that are most affected by the change of amplitude at the frequency f(x) can be defined as "hot spots" for the corresponding biological function x. The rationale behind these predictions is explained in this chapter.

Knowing the characteristic frequency of a particular protein or the function of a DNA fragment, it is possible to predict which amino acids or nucleotides in the sequence predominantly contribute to the frequency and consequently to the observed function [24, 25]. Since the characteristic frequency f(x) correlates with the biological function x, the positions of the amino acids that are most affected by the change of amplitude at the frequency f(x) can be defined as "hot spots" for the corresponding biological function x. The strategy for this prediction includes the following steps:

1. Determining the characteristic unique frequency for the specific biological function by multiple cross-spectral analysis for the group of sequences with the corresponding biological function.

 ©1997, Birkhäuser Verlag Basel

2. Altering the amplitude at this characteristic frequency in the particular numerical spectrum. The criterion used for identifying the critical character istic frequency change is the minimum number of hot spot amino acids or nucleotides that are least sensitive to further changes in the amplitude of the characteristic frequency.
3. Deriving a numerical sequence from the modified spectrum using DFT.

It is known [43] that change in amplitude at one frequency in the spectrum causes changes at each point in the numerical sequence. Thus a new numerical series is obtained where each point is different from those in the original series. Determining the amino acid corresponding to each element of this new numerical sequence can then be achieved using tabulated values of EIIP or other appropriate amino acid coefficient parameters. The amino acids or nucleotides in the new sequence which differ from the original ones reside at the points mostly contributing to the frequency f(x). These hot spots are related to this frequency and to the corresponding biological function. The procedures described have already been applied to a number of examples: previous studies with IL-2 [29], SV40 enhancer [28], TNFs [24], Ha-*ras* p21 oncogene products [32, 33, 64], glucagons [25], hemoglobins, myoglobins and lysozymes [31]. All these studies have documented evidence that such predicted amino acids denote residues crucial for protein function. Furthermore, in the examples studied to date [25, 31, 32, 33], these amino acids are found to be spatially clustered in the protein tertiary structure and to be positioned in and around the protein active site.

4.2 Example of heme-containing proteins

In a number of examples including hemoglobin, myoglobin, lysozyme and cytochrome C, it was found that amino acids which mostly contribute to the characteristic frequency are clustered around the active site of the protein. Since these amino acids are at the locations where the resonant characteristic frequency signal is dominant, their cluster may define specific sites in the 3D structure of a protein which act as resonators for the characteristic frequencies. These resonators are located just around the cleft of the active site and thus can be considered as resonant boxes which provide optimum conditions for resonant energy transfer at the frequency characteristic of the protein function. As a consequence, the active protein conformation can be predicted to provide optimal resonant conditions for a particular frequency characteristic, thus dictating the specificity of biorecognition between proteins and their targets and the subsequent transfer of energy which both define the functional selectivity of the biomolecular interactions At this stage, the relationship between the RRM and 3D protein structure is still not clear. Consequently, it would appear that the proteins are folded in such a way that, where the characteristic frequency is strongest, amino acids are clustered together and positioned around the active sites. This has implications for the overall structure of proteins.

This study describes the further extension of the RRM for the analysing and predicting protein-protein and protein-DNA structure-function dependencies. Using sperm whale myoglobin (SWM), human hemoglobin, cytochrome C and hen egg white lysozyme as model protein examples, sets of predicted amino acids, or so called hot spots, have been identified within the tertiary structure. For each protein, it was found that the predicted hot spots distributed along the primary sequence are spatially clustered around the active site. The amino acids identified did not correspond to the amino acid residues involved in the chemical reaction sites of these proteins. It is thus proposed that the RRM approach enables the identification of amino acid residues important for the creation of the molecular structure around the catalytic active site and the associated physical field conditions required for biorecognition, docking of the specific substrate and full biological activity to occur.

Myoglobins and hemoglobins

In the first instance, the RRM approach was applied to myoglobins and hemoglobins as two functionally and structurally well defined groups of proteins. Both groups are involved in oxygen storage, but their functions are specific to different cells: hemoglobins to erythrocytes and myoglobins to skeletal muscle cells. It was therefore important to ascertain whether a different characteristic RRM frequency was identifiable for each of these two groups of proteins which would characterise their specific cellular function and activity. As shown in Chapter 2, for 49 myoglobins of different origins only one common frequency ($f = 0.0820 + 0.004$ and signal-to-noise ratio of 128:1) was found to characterise their common function, and a different common frequency was found (f = 0.0234 + 0.008 with signal-to-noise ratio of 119:1) for the group of 187 hemoglobins. The cross-spectral functions of myoglobins and hemoglobins are represented in Figures 2.8 and 2.9, respectively.

Once the characteristic frequency has been determined, it is possible to identify amino acids in the sequence of any individual protein which are related to this frequency using procedures outlined in Section 4.1.

SWM is a single polypeptide chain comprising 153 amino acid residues. Compared with many globular proteins, myoglobins are unique in terms of their α-helix content. X-ray crystallographic studies [65] indicate that approximately 75% of the amino acid residues are arranged in segments of right-handed α-helices. These helical arrangements lead to a compact and roughly spheroidal molecule with dimensions 45 × 35 × 5 Å.

The hot spot prediction for SWM identified amino acid residues at the positions Gly 25, Phe 33, Phe 46, Gly 65 and Gly 124 [31]. The spatial disposition of these hot spots is shown in Figure 4.1 using the 3D structure of the SWM molecule. Within the protein molecule, each hot spot is labelled and highlighted with a shaded CPK surface (ball presentation of each atom), with the heme moiety shown. Even though the hot spot residues are distributed throughout the primary sequence, it can be observed from the 3D presentation that the hot spots (except Gly 125) are in close spatial proximity in the folded protein. It is also noteworthy that these hot spots do not include the two amino acids (His 64, His 93) which form the chemical active site together with the heme moiety; rather the hot spot amino acids are clustered over the active sites. Similar arrangements were found for predicted hot spots for myoglobins originating from different species. This result indicates that the amino acid residues which characterise the hot spots are important for the formation of the biologically active molecular structure with the consequent appropriate physical field conditions around the active site [25, 31].

To evaluate the generality of this concept, similar analyses were carried out with hemoglobins as another group of heme-containing proteins. In contrast to myoglobin, hemoglobin is made up of four polypeptide chains held together by non-covalent interactions. With human hemoglobin A, for example, there are two distinct polypeptides, namely the α (142 amino acids) and β (146 amino acids) chains. The two α and two β chains are packed together in a tetrahedral array to form an almost spherical molecule with a diameter of 55 Å [65]. The predicted hot spot

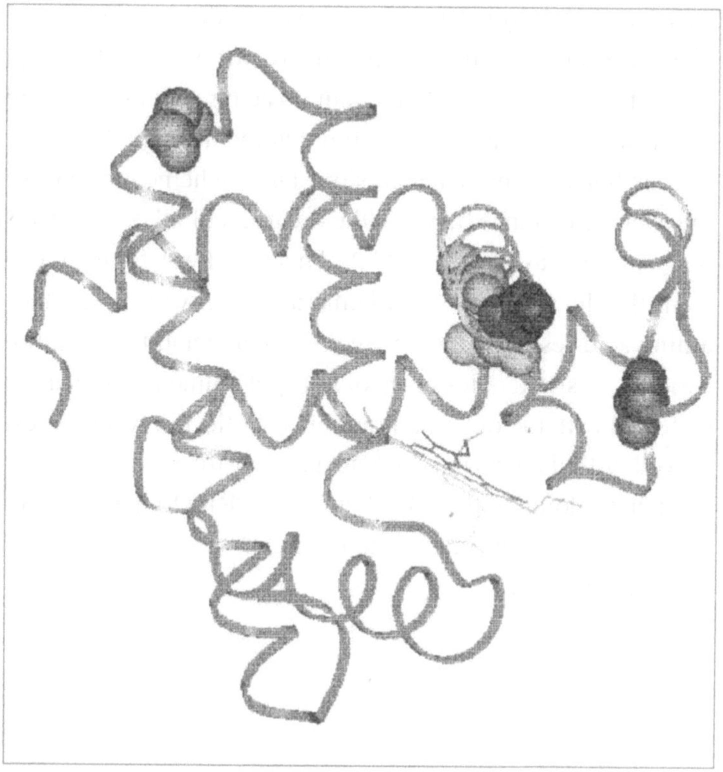

Figure 4.1 Three-dimensional structure of SWM in backbone presentation
Predicted hot spot amino acids are highlighted with shaded CPK surfaces and are clustered in a space around the cleft of the active site.

amino acid residue positions in the β subunit of human hemoglobin are Phe 42, Phe 45, Gly 64 and Gly 69, and in the α subunit Gly 18, Gly 22, Phe 36, Phe 43 and Gly 59 [25, 31]. The spatial position of these hot spots within the 3D α-carbon backbone is shown in Figures 4.2 (β subunit) and 4.3 (α subunit). The hot spot amino acids are highlighted with CPK surfaces. These 3D presentations show clearly how very close spatially the amino acids which constitute the hot spots are to each other. Similar observations were made for other hemoglobins evaluated in an analogous manner. It is important to note that in the case of hemoglobins, as in the case of myoglobins, the predicted hot spots are not heme-linked amino acids (His 63, His 92). Even though the hot spot amino acids do not correspond to the chemical active site residues, they are all spatially clustered around the heme. The statistical significance of the proximity of the hot spot amino acid residues to the active site was determined by calculating the average distance and corresponding standard deviations from the heme group to the α carbon of each hot spot amino acid residue and to the α carbon of all other amino acid residues separately. For example, it was found with α human hemoglobin that the average distance between the α carbon of each hot spot amino acid residue and the heme group was 13.89 ± 0.85 Å, while the average distance between the α carbon of all other amino acid residues and the heme group was 15.95 ± 0.23 Å. This result was found to be significantly different from the distances between heme as an active site and the α carbon of hot spot amino acid residues and between the heme and the α carbon of other amino acid residues. A similar clustering of hot spot amino acids over the heme moiety was also found with the other hemoglobins. Collectively, these results are completely in accord with the previous conclusion that for myoglobin the hot spot amino acids (which predominantly contribute to the RRM characteristic frequency) are clustered over the heme active site in the 3D structure of these three families of heme proteins [25, 31].

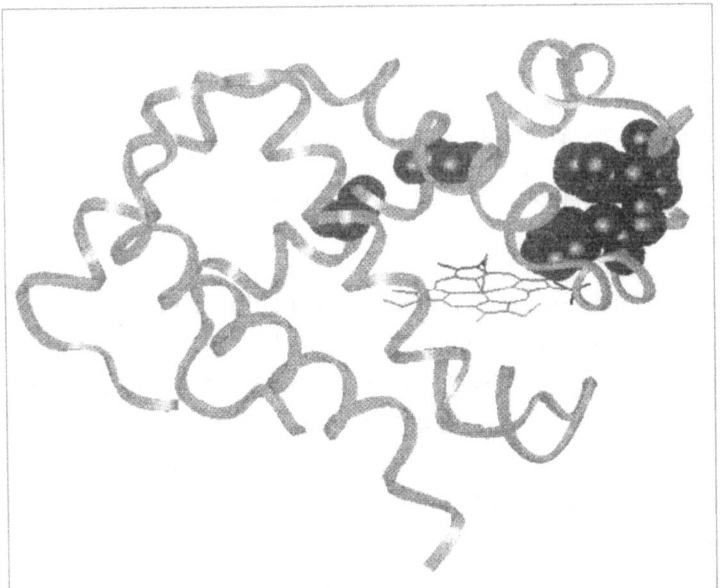

Figure 4.2 Human hemoglobin β subunit 3D structure in backbone presentation

Predicted hot spot amino acids are highlighted with shaded CPK surfaces and are clustered in a space around the cleft of the active site.

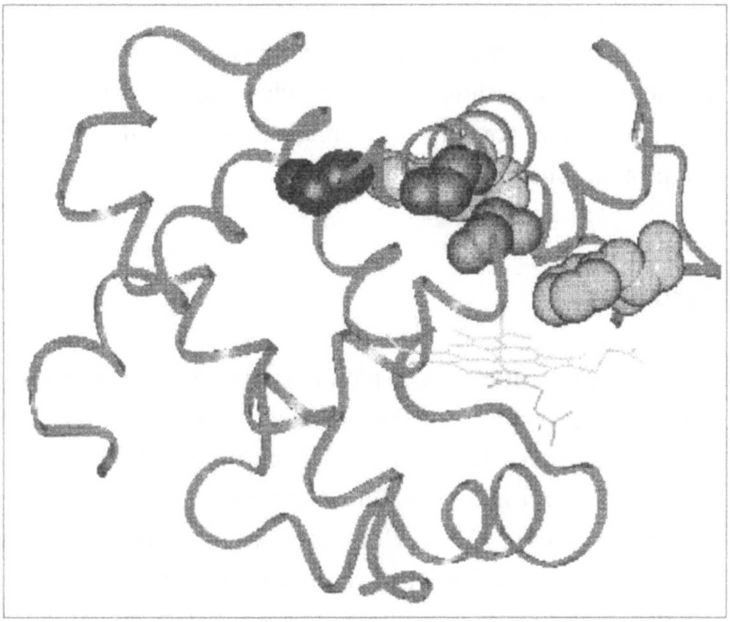

Figure 4.3 Human hemoglobin a subunit 3D structure in backbone presentation

Predicted hot spot amino acids are highlighted with shaded VDW surfaces and are clustered in a space around the cleft of the active site.

Cytochrome C

The prediction of hot spots with typical results is now described for the case of cytochrome C proteins. Their common biological function is electron transfer through excitation of the heme group, which is positioned in the cleft of the active site within the 3D protein structure [66]. Cross-spectral analysis of 38 cytochrome C proteins from different origins revealed a common frequency component at f = 0.476 + 0.004 with S/N = 252 (Fig. 2.5e). Consequently, the hot spot amino acids in tuna heart cytochrome C were determined at positions 41, 45, 56 and 77. Although these amino acids are not sequentially linked, they form a spatial cluster in the 3D protein structure positioned around the cleft of the protein active site (Fig. 4.4) [25]. Numerous other studies found that hot spot amino acids were clustered around the active site of the protein as well [25, 31, 32, 33]. Since hot spots are the amino acids whose resonant characteristic frequency signal is dominant, the cluster of hot spots may define specific sites in the 3D structure of a protein which act as a resonator for the characteristic frequency. This resonator is located just around the cleft of the active site and thus can be considered as a resonant box which provides optimum conditions for transfer of resonant energy at the frequency characteristic of the protein function.

Lysozymes

To determine wether such spatial arrangements of hot spots exist only in heme-containing proteins, we have analysed lysozymes as an example of a protein family without a prosthetic group. The common RRM frequency for 15 lysozymes from different origins was found at f = 0.3281 + 0.004 (with signal-to-noise ratio of 124:1) (Fig. 2.5e). Hot spot analysis was then carried out with hen egg-white lysozyme (HEWL) as representative of the lysozyme family. HEWL is a small enzyme made up of a single polypeptide chain of 129 amino acids. X-ray crystallographic studies have revealed that the HEWL molecule is approximate-

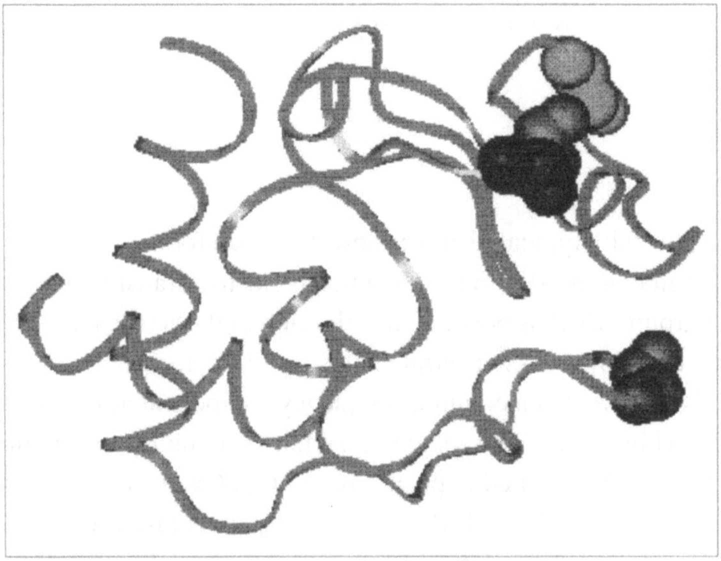

Figure 4.4 Cytochrome C (tuna heart) 3D structure in backbone presentation

Predicted hot spot amino acids are highlighted with shaded VDW surfaces and are clustered in a space around the cleft of the active site.

ly ellipsoidal with dimensions 45 × 30 × 30 Å and divided into two domains by a deep cleft located on one side of the molecule. X-ray cystallographic and solution studies on enzyme-substrate and enzyme-inhibitor complexes have identified the cleft as the active site for the degradation of gram-negative bacterial cell wall polysaccharides [65].

The hot spots corresponding to the characteristic frequency of lysozyme (f = 0.3281) were found to be at positions Phe 3, Gly 4, Gly 16, Gly 22, Ser 24, Cys 30,

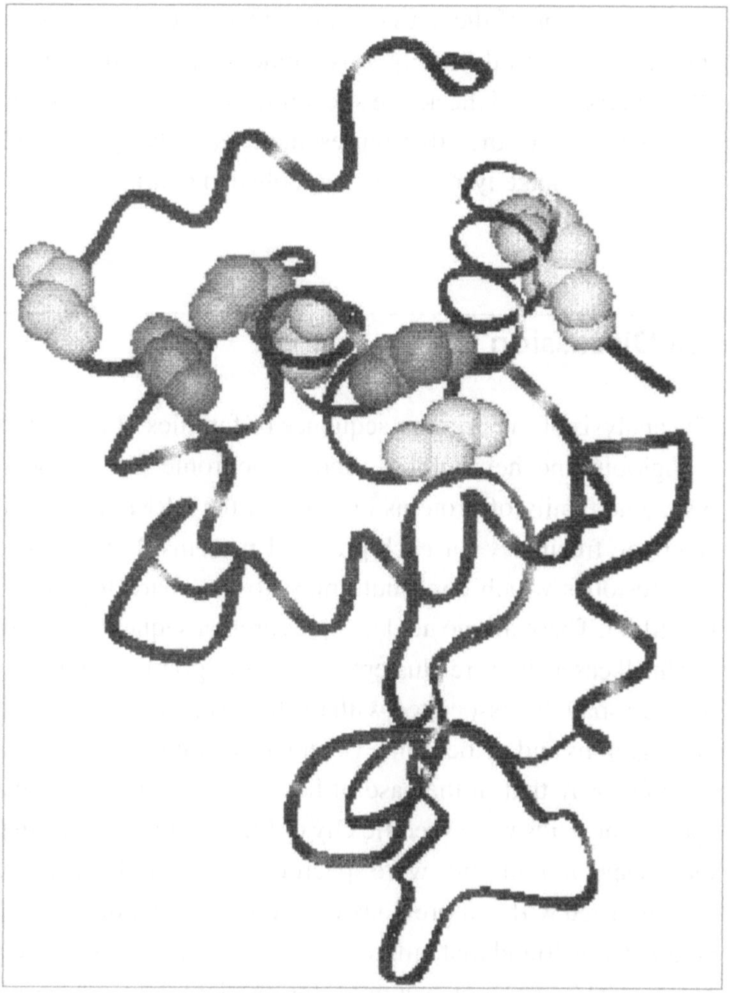

Figure 4.5 HEWL 3D structure in backbone presentation

Predicted hot spot amino acids are highlighted with shaded VDW surfaces and are clustered in a space around the cleft of the active site.

Ser 36 and Gly 117 in the HEWL sequence. The predicted hot spots are presented in the 3D backbone structure of HEWL as complete amino acids with side groups highlighted with CPK surfaces in Figure 4.5. In the case of lysozymes, the active site is formed by amino acid residues which participate in three different types of interactions. These comprise hydrogen bonding (Asp 101, Trp 62, Trp 63), van der Waals interactions (trp 57) with polysaccharide substrates and catalytic active sites (Asp 52, Glu 35). It is important to note once again that the predicted hot spots in HEWL do not correspond to the amino acids at the chemical catalytic site. The relative orientation of the amino acid residues identified as hot spots does not correspond to sequentially contiguous amino acids, but these residues are nevertheless all clustered around the active site amino acids in the catalytic cleft. Similar results were obtained for the other representatives of the lysozyme family (including human and T4 phage lysozymes, unpublished data).

4.3 Discussion

The analysis of the primary sequence of families of the heme-containing proteins, myoglobin and hemoglobin, and cytochrome C and the non-heme-containing lysozyme family of proteins in terms of the RRM has established a distinct characteristic frequency for each group of proteins. Furthermore, the hot spot amino acid residues which contribute most to the characteristic frequencies have been identified. These amino acid residues are not sequentially contiguous amino acids, but in all cases they are clustered in the topographic region over the active site; i.e., they are spatially associated with the heme catalytic site in the case of myoglobins, hemoglobins and cytochrome C, and the catalytic cleft in the case of lysozymes. It is noteworthy that in the case of heme proteins the majority of the predicted hot spot amino acids were specific Gly or Phe residues, whilst in the case of lysozymes the hot spot amino acids were specific Gly, Phe and Ser residues. These results can be contrasted with our previous work on hot spot amino acids within other proteins where it was found that amino acids other than Gly and Phe were predicted as the hot spot residues. For example, the predominant amino acids are specific Met residues in the functional cluster in the case of IL-2 [29], specific Ser and Cys residues in the case of epidermal growth factor (EGF) and specific Thr residues in the case of glucagon [25]. Furthermore, different amino acids and different amino acid positions are predicted as hot spot residues when the analyses are carried out with the same protein but using a different characteristic frequency associated with a different functional property [29, 64]. Thus, when these results are considered in conjunction with the mathematical basis of the RRM in terms of signal analysis theory, we can conclude that identification of hot spot amino acids using the RRM depends on the structure of the whole protein and the characteristic frequency rather than on the EIIP value for a particular amino acid.

In all proteins studied here, the hot spot amino acids do not correspond to amino acid residues which are directly involved in the chemical reaction between the protein and its ligand (substrate). The significance of previous RRM investigations can now be evaluated in the light of these results.

As described above, RRM characteristic frequencies have been proposed to represent the oscillations of a physical field which are responsible for information transfer between interacting biomacromolecules. If this field is generated by charge movement through the backbone of the biomacromolecule as described above, then the frequency range of such a field will be dependent on the charge velocity. The RRM model is based on these concepts and represents biological function as a resonant energy transfer from one biomacromolecule to another. As a consequence, it can be concluded that RRM characteristic frequencies are a relevant parameter for biorecognition between biomacromolecules and that they are important in describing the resonant energy transfer between interacting biomacromolecules and their substrates or targets. The hot spots are the amino acid residues in the sequence which influence the characteristic frequency mostly. Therefore, they represent amino acid residue positions in the protein or polypeptide where the signal of the resonant characteristic frequency is dominant. It has been found that these hot spots, in all the proteins analysed, are clustered in the space around the active site. These hot spots may thus define specific amino acid residue sites in the 3D structure of a protein which act as a resonator for the characteristic frequency. As this resonator is located just around the chemical catalytic site, the protein molecule can be considered as a type of "resonant box", in which the space around the active site provides optimum resonant characteristics for the frequency which corresponds to its unique function. As a consequence, the active protein conformation can be predicted to provide optimal resonant conditions for a particular frequency characteristic, thus dictating the specificity for protein biorecognition with its target and the subsequent energy transfer which defines the functional selectivity of the biomolecular interaction.

5 Prediction of Functionally Active Sites and Functional Mutations

Some parts of a protein molecule can be changed or deleted without any significant change in protein function, whereas sometimes even a single amino acid change can result in complete functional change or destruction. Thus, while the entire protein is not involved in molecular interactions (biological function), some parts of it, or even single amino acids, are more crucial than others for protein function. These functionally more important parts or amino acids are called the active sites. It would be most desirable to have a predictive method to locate active sites in the protein. If this were possible, it could form a basis for rationally engineering protein mutations that could lead to desired functional changes. This chapter describes applications of the RRM to predicting functionally active sites and functional mutations.

5.1 Hot spot predictions in DNA: example of the SV40 enhancer

Enhancers are DNA sequences responsible for regulating gene expression. Hot spot predictions were tested on the example of the SV40 enhancer as a model DNA system. The SV40 enhancer was chosen because of the availability of data on the activities of systematically obtained mutants. Predicted and experimentally obtained results have since been compared, and significant agreement has been achieved.

Regulating gene expression still remains a great mystery of life. However, some facts are known, and further investigation might elucidate this complex process and even allow us to control it. One approach to this kind of research is to evaluate the most important base positions in elements that control gene expression

 ©1997, Birkhäuser Verlag Basel

via systematic mutagenesis. One such study was carried out [61] for the SV40 enhancer, as it is the best-analysed enhancer element [67]. Enhancers, DNA regulatory sequences, have been identified in many viral genomes as well as in controlling regions of some eukaryotic genes [68]. They enhance transcription independently of position, orientation and distance and are able to induce the expression of heterologous genes [68]. Most of them show host or tissue specificity; i.e., they are active only in particular cells [68]. This strongly indicates a requirement for transacting cellular components. These regulatory sequences are of great interest, and their systematic modification could help to explain the mechanism of their action. However, these kinds of experiments are long and laborious. Thus, a basis for the theoretical prediction of important nucleotide positions might be helpful.

It is very hard to test the effect of single-point mutations on protein function, and consequently it is difficult to test a hot spot prediction experimentally. Thus, there is not enough experimental data on all possible mutations in a particular sequence. For example, if a sequence has 200 amino acids, it is possible to make $200 \times 19 = 3800$ different single-point mutations within this sequence. With DNA, the situation is slightly easier: for a 200-nucleotide sequence, it is possible to make $200 \times 3 = 600$ different single-point mutations. Since this is a vast undertaking, the value of being able to predict hot spots, even with only a small degree of certainty, is obvious.

The common procedure is to analyse the activity of a few, usually randomly chosen, mutants. The question here is, What would happen with other mutants not obtained in that particular experiment? That is, would they be more or less relevant for the function of the tested molecule? With the SV40 enhancer, however, the situation is different. A systematic mutagenesis of the SV40 enhancer was performed in order to determine the DNA sequence essential for the activity of this "prototypal" enhancer element [67]. With deletions and mutations constructed throughout the enhancer region, and by measuring the level of transcription in vivo, it was concluded that the SV40 enhancer is spread over a region of about 100 bp and that it is composed of at least two domains. Domains A and B are located inside the 72 bp, while the putative third domain lies in the 5′ flanking region. These regions approximately include nucleotides 185-220, 225-257 and 267-276 (with respect to the nomenclature in [67]).

Although these experiments confirmed that the "core" sequence $G^{AAA}_{TTT}GT$, also identified in the other enhancer sequence, is the most important, the existence of several additional sequence elements critical for enhancer activity were reported

[67, 68]. It is thought that each of these multiple elements is recognised by specific components of the transcriptional machinery and that the simultaneous functioning of all these elements maximises enhancer activity [2, 23].

The first step in this research was to determine the enhancer characteristic frequency [27, 28]. The RRM was applied to the following enhancers: SV40 (143 nucleotides), Moloney sarcoma virus (MSV 144 nucleotides), polyoma virus (249 nucleotides), human immunoglobulin heavy-chain locus enhancer (583 nucleotides), immunoglobulin enhancer from the unarranged IgM constant region (313 nucleotides), bovine papilloma virus (BPV, 64 nucleotides), mouse mammary tumor virus (MMTV, 153 nucleotides), mouse immunoglobulin K gene lymphocyte-specific enhancer fragment (479 nucleotides), mouse immunoglobulin heavy-chain locus enhancer region (298 nucleotides), adenovirus enhancer (160 nucleotides) and human β-interferon gene enhancer (41 nucleotides).

The Fourier transforms of the sequences of the EIIP values corresponding to the enhancer nucleotide sequences were calculated with a 1024-point transform. Since the sequences were all shorter than 1024 nucleotides, the signal was padded with zeros. The frequency associated with any peak in the resulting transform is $n/1024$ where n is the point in the transform where the peak occurs. The common characteristic frequency for all enhancers analysed was found to be at $f = 0.0488$ with $S/N = 467$ [27, 28]. The sequences considered differ in length by 41-583 nucleotides. The error in defining the common frequency was calculated according to the shortest sequence (41 nucleotides) and equal to ± 0.024. This value equals the resolution of the spectrum of the shortest sequence. The existence of the common frequency implies that all the analysed enhancers have only one frequency component in common. Considering their common biological function, i.e. transcriptional enhancement, this characteristic frequency would appear to be related to the enhancer function. Figure 5.1 illustrates the cross-spectra for the group of the enhancers listed above.

The next step was to observe the single spectrum of the SV40 enhancer. The amplitude at the enhancer characteristic frequency, $f = 0.0488$, was altered by different degrees, and the resulting changes in the sequence were observed. Thus, if there exists a correlation between frequency f and the biological function of the enhancer, then using the inverse procedure, decreasing or increasing the amplitude at this frequency by some factor K, would result in changes in the nucleotide sequence that should be related to the enhancer's functional sites. Similarly, base positions irrelevant to the enhancer activity were obtained by removing the amplitude at the char-

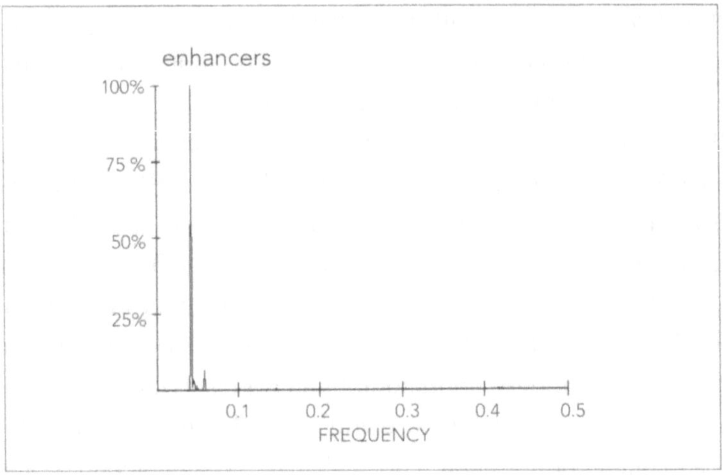

Figure 5.1 Multiple cross-spectral function of enhancers
Characteristic frequency is at f = 0.0488 ± 0.024.

acteristic frequency f in the single spectrum of the SV40 enhancer, and by noting base positions which remained unaltered.

The next question to be determined is whether these base changes, obtained theoretically, are related to the positions of mutations which affect enhancer activity *in vivo*. To answer this question, two approaches are used. First, the amplitude values at the characteristic enhancer frequency are observed for different mutants with reported activity. Four mutants whose base alternations affect the enhancer activity dramatically are chosen. Comparing the amplitude values between the SV40 wild-type enhancer and chosen mutants at the characteristic frequency, it can be seen (Fig. 5.2) that mutants which increase transcriptional activity have a higher amplitude and S/N at the characteristic frequency and vice versa.

The second approach is to predict base positions which should be changed for a defined efficiency of the activity of the enhancer. The amplitude at the characteristic frequency in the spectrum of the SV40 enhancer was altered by a factor K, so that altered amplitude was equal to the original amplitude times K, divided by 100.

After altering the amplitude, the inverse Fourier transformation of the altered spectrum yielded a new sequence. For each position in the sequence, the maximal factor for decreasing the amplitude and the minimal factor for increasing it required to bring about a different nucleotide at that position in the sequence were determined. Figure 5.3 represents a plot of these factors versus the nucleotide se-

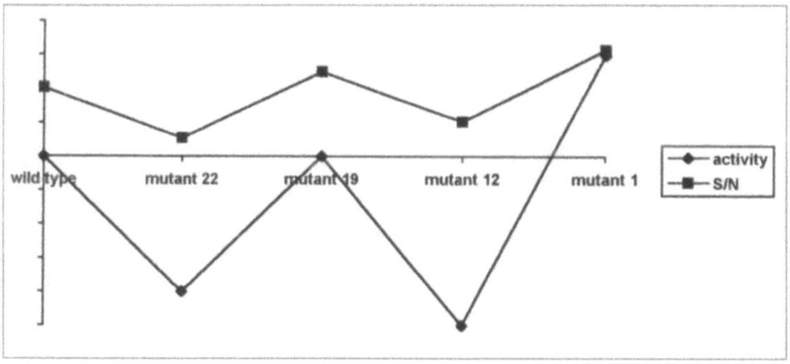

Figure 5.2 Biological activities and S/N at the enhancer characteristic frequency for different SV40 enhancer mutants

(a) Theoretically obtained contribution of each nucleotide to the efficiency of the activity of the SV40 enhancer; the abscissa represents the SV40 enhancer nucleotide sequence, while the ordinate is the reciprocal of factor K, defined in Chapter 4.

quence and a comparison of these theoretical results with an activity obtained experimentally. Since this prediction was carried out for each nucleotide, while experiments were performed on triplets, differences that existed at the edges of the domains were acceptable. However, the similarity is obvious. This means that theoretically it is possible using the RRM approach to read the fine structure of the contribution of each particular point in the sequence to the biological function of a given molecule.

All experiments involving systematic mutagenesis require laborious experimental work, especially when one cannot predict the kind of changes that will be relevant for the function of a particular molecule. It is very hard to test every single-point mutation in even one small molecule. Furthermore, the question still remains, Is that change the most important one, or is there, in a neighbouring sequence, a point which would interfere more with function?

The RRM obviates some of this tedious experimental work. Mutagenesis experiments performed for the SV40 enhancer element allowed us to test the method using a very detailed experimental account [1]. Thus, when the theoretical predictions were compared with the experimental data, the functional importance of each point turned out to have been predicted correctly. Because these theoretical predictions are comparable with the experimental results, the same method can be applied to other macromolecules, both DNA sequences and proteins. The result of this kind of analysis might be of great help in designing further experimental work.

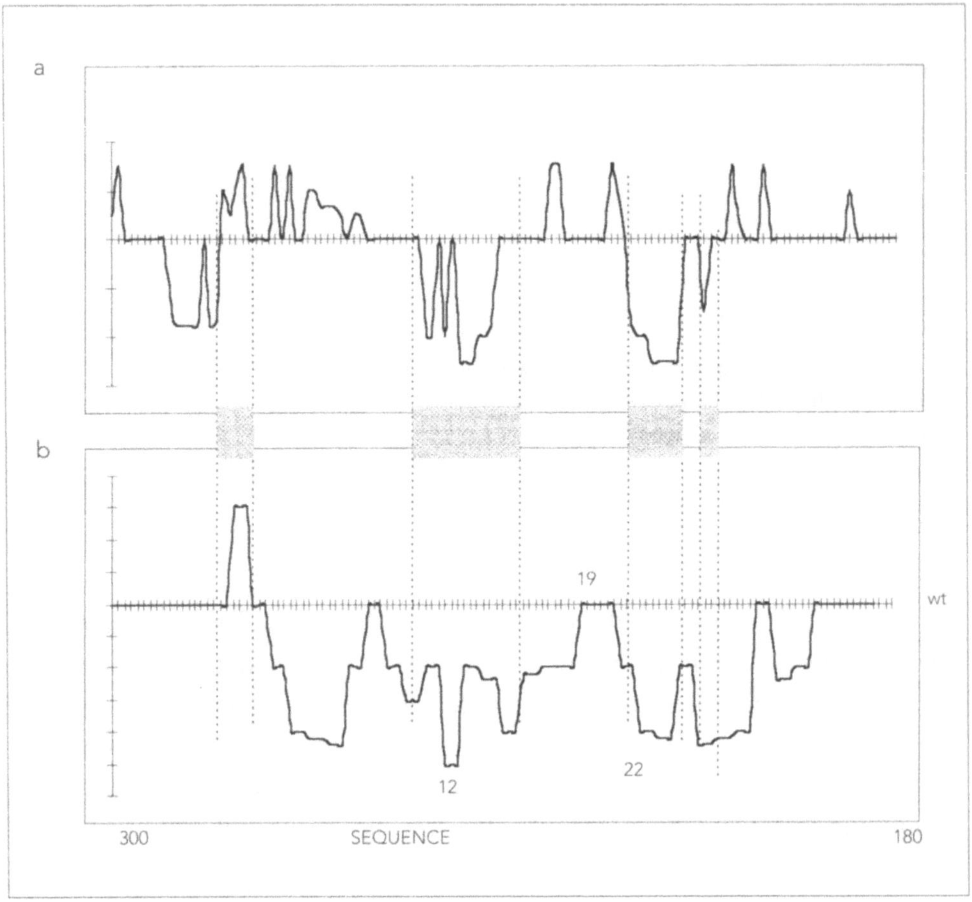

Figure 5.3 Schematic presentation of the comparison between experimental transcription level and theoretical predictions

The abscissa represents the nucleotide sequence of the SV40 enhancer. (a) Theoretically obtained contribution of each nucleotide to the efficiency of the calculated function of the SV40 enhancer, based on the value of factor K, defined in Chapter 4. As K approaches zero, the importance of the nucleotide is higher. For K = 0, the importance is zero as well. (b) Level of transcription in vivo for 34 mutants. Each mutant has a mutation in a triplet. The number of the mutant then represents the triplet number which was altered, beginning at nucleotide position 279, continuing to position 178 (according to the nomenclature in [67]). Dotted lines and shaded areas represent sequence regions where complete agreement between prediction and experimental result was found.

5.2 Hot spot predictions in proteins: example of IL-2

IL-2 is a protein that has a variety of effects on lymphocytes, particularly in promoting growth. Hot spot predictions have been tested on IL-2 as a model protein system because its active domains have already been determined experimentally. Predicted and experimentally obtained results were compared, and significant agreement was obtained.

IL-2 affects lymphocytes in a number of ways: promoting proliferation of activated T cells, production of lymphokines, and growth and differentiation of activated B cells, cytotoxic T cells, NK cells and lymphokine activated killer (LAK) cells [29]. It has also been shown that administration of IL-2 and autologous LAK cells is potentially useful in the treatment of neoplastic diseases [29].

It is, however, interesting to see whether amino acids that are important for specific activities of the IL-2 molecule can be located within the molecule. Different amino acid residues on the surface of the IL-2 molecule bind to α and β chains of the IL-2 receptor [29], and it appears that different distributions of IL-2 α (55 kD) and β (70 kD) receptors are important in different IL-2 activities. There is now evidence that IL-2R-β mediates the initial phase of the activation of resting NK and LAK cells by IL-2. Some studies have indicated that carboxy and amino termini of IL-2 are important for maximal cytotoxic T-lymphocyte line activity, and that both terminal regions and amino acids 36-46 or 30-60 of human IL-2 are important for IL-2 proliferative activity [29, 69].

The primary structures of IL-2 molecules, human murine and bovine, were analysed, and three prominent peaks were found in their multiple cross-spectrum: $f = 0.293 \pm 0.008$, 0.308 ± 0.008 and 0.027 ± 0.008 [29, 70]. These frequency components were compared with the growth-promoting characteristics of other growth factors and lymphokines. The most prominent peak in IL-2 is at $f = 0.027 \pm 0.008$. Several lymphokine groups were found to have a characteristic frequency within this frequency range. For example, the prominent characteristic of thymosines α1 and α7, TNFs, IFNs-β and M-CSF (CSF-1) is at 0.031 ± 0.016, which is within the same range as for IL-2. The common functional feature of all these factors, including IL-2, is their antitumor and immunoregulatory activity. Another prominent characteristic of IL-2 is obtained at the frequency $f = 0.293 \pm 0.008$. This frequency component was found to be common in several groups of growth-stimulating factors, including EGFs (epidermal growth factors), proliferins, NGFs (nerve growth factors), growth hormones, IL-1 (interleukin-1), IL-4, IL-6, ubiquitins, EPA and EPA homologue products of the growth-responsive gene [70]. Hence, this frequency could be related to general growth-promoting activity. There are only a few groups of growth factors that do not have a significant amplitude in this frequency range: FGFs (fibroblast growth factors), insulin and IGFs (insulin-like growth factors), TGFs (transforming growth factors) α and β. We consider the frequency range 0.293 ± 0.016 to be essential for growth-promoting activity. In addition, IL-2 and its receptors do have a common characteristic frequency at 0.308 ± 0.008. As this frequency is common to both IL-2 and its receptors, it can be considered as relevant for IL-2-receptor recognition.

We propose that the above three frequency domains correspond to the following functions:

1. growth stimulation ($f = 0.293 \pm 0.008$)
2. antitumor and immune system control ($f = 0.031 \pm 0.008$)
3. IL-2 - receptor recognition ($f = 0.308 \pm 0.008$)

If a correlation between these frequency domains and the biological function of IL-2 exists, we can expect that hot spots identified by the RRM will correspond to the amino acids that are experimentally determined as functionally important.

The minimal change of the amplitude value at a certain frequency resulting in an amino acid change appears to be the most significant for this frequency and, consequently, for the corresponding biological function. The results are schematically presented and compared with the predicted receptor binding site in IL-2 in Figure 5.4. Six out of eight predicted hot spots are within the potential receptor binding site. Three of the amino acids in the potential receptor binding site are the same as the first three hot spots related to the main growth-promoting frequency (0.293): 39, 44 and 46. The probability of obtaining this result by chance is 0.1%. The fourth predicted hot spot, 29, is in the N-terminal region of IL-2, essential for IL-2 activity [70]. The hot spot at position 27 was the most important for frequencies 0.031 and 0.308, and this amino acid is also within the important functional region at the N-terminus, while the hot spot at position 42 is within the potential receptor binding region. In addition, the hot spots at positions 44 and 58 were experimentally found to be important for IL-2 function. The hot spot at position 37 relates to the frequency 0.308, and it is also within the potential receptor binding site.

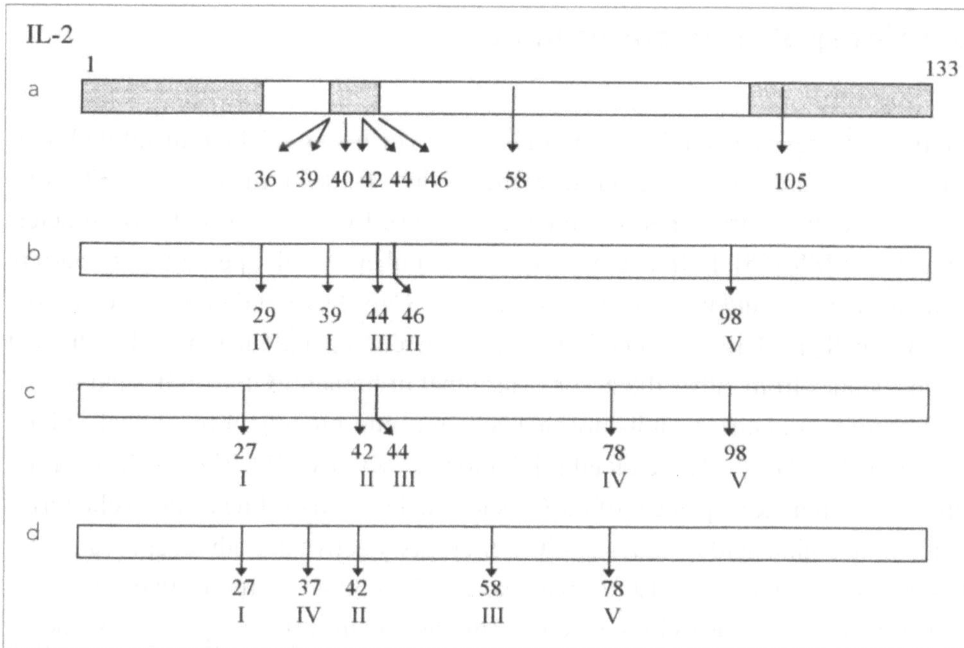

Figure 5.4

(a) Summary of reported IL-2 structural changes experimentally proven to be important for IL-2 function. Amino acids 42, 44, 58 and 105 and a region from amino acids 36 to 46 (36, 39, 40, 42, 44, 46) are likely to bind IL-2 receptors. Shaded areas represent deletions that completely destroy IL-2 activity [32]. (b) Theoretically predicted hot spot amino acids related to characteristic frequency of the the main growth factors: 0.293 ± 0.008. Numbers under each hot spot amino acid denote the degree of its contribution to the characteristic frequency (I → the most contributing, V → the least contributing). (c) Theoretically predicted hot spot amino acids related to the characteristic frequency of antitumor factors: 0.031 ± 0.008. (d) Theoretically predicted hot spot amino acids related to the IL-2–IL-2 receptor frequency: 0.308 ± 0.008. Most of the predicted hot spot amino acids fall within the active amino acids or regions obtained experimentally.

Only hot spots at positions 78 and 98 match neither the results for mutations or deletions related to the IL-2 function, nor the theoretically predicted IL-2 receptor binding site [70]. However, no experimental evidence has yet shown whether these amino acids are important for IL-2 function. It should also be noted that some of the predicted hot spots for different frequencies are the same. These spots are concentrated mainly in the potential receptor binding site. It is possible that this coincidence reflects the realisation of multiple IL-2 functions through interaction with the same receptor. Predicted hot spots in bovine, human and mouse IL-2 are also within conserved regions [29].

5.3 Hot spot predictions in EGF

EGF is a 53-residue, single-chain polypeptide consisting of three disulfide bonds which define three looped regions from the residues 1-20(A), 14-31(B) and 32-53(C) [71]. The tertiary structures of human and mouse EGF have recently been determined by NMR [65]. EGF is mitogenic for a number of cell types and has a variety of actions *in vitro* and *vivo*, such as to stimulate RNA, DNA and protein production, enhance cell proliferation and inhibit gastric secretion [71]. The removal of the first three residues from either the N- or C-terminal of human EGF (hEGF) and mouse EGF (mEGF) did not result in any change in receptor binding or biological activity of EGF [71]. EGF-(1-47) retained full biological activity [71]. The results overlapping with synthetic sequences of mEGF showed that both the linear and cyclic forms containing amino acid residues 21-30 were weakly active in both receptor and biological assays (0.003% of EGF), whereas residues 32-48 were inactive. EGF-(15-53) proved as active as EGF in receptor binding and in stimulating DNA synthesis. In addition, fragments 42-53 and 32-53 were incapable of inducing DNA synthesis. EGF-(1-48) retained full receptor binding and mitogenic activity, EGF-(1-47) decreased three- to sevenfold, and EGF-(1-46) by 100-fold, while EGF-(1-45) and EGF-(1-21) were inactive. Using site-directed mutagenesis mutation of Leu 47 to His 47 resulted in a marked reduction (7% of wild type) in receptor binding activity. The structure-function relationship of EGF remains still to be defined, despite substantial efforts on the part of different groups to elucidate it.

Because EGF has clinical potential for wound healing and antitumor treatment, an EGF agonist or antagonist could be an important therapeutic tool in the treatment of certain types of trauma and malignancies. Understanding the conformational and physiodynamic properties of the EGF molecule is basic to designing an antagonist or agonist. Thus, the RRM was applied to determining the most crucial amino acid residues that required for maintaining an active EGF structure.

Cross-spectral analysis of hEGF, rat (rEGF), guinea pig (gEGF) and mEGF epidermal growth factors revealed two prominent characteristic frequencies at $f = 0.062 \pm 0.02$ and 0.288 ± 0.02 (Fig. 5.5a). These two informational frequencies are considered to be relevant for biological recognition between receptor and hormone. Consequently, when this spectrum was compared with the hEGF receptor, only one common frequency ($f = 0.062$) resulted (Fig. 5.5b). The second frequency ($f = 0.288$) belongs to the already identified frequency range 0.28-0.30 found to be com-

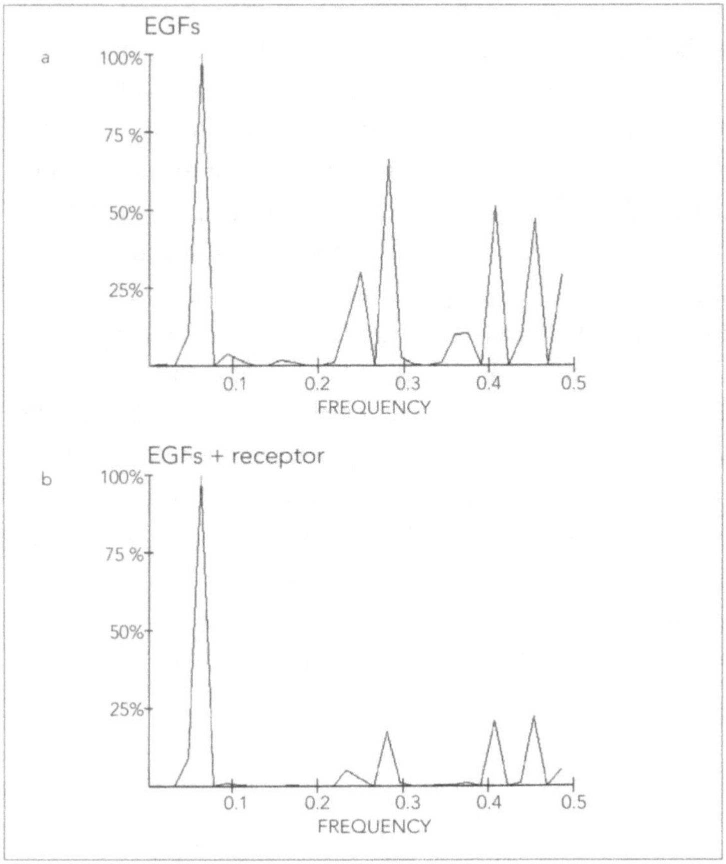

Figure 5.5
(a) Multiple cross-spectral function of four EGFs from different origins (human, rat, mouse and guinea pig). Two prominent peaks are at frequencies 0.062 ± 0.02 and 0.288 + 0.02. (b) Multiple cross-spectral function of four EGFs and a human EGF receptor. Only one prominent peak appears at the frequency 0.062.

mon to most growth factors and associated with general growth-promoting function [25, 69]. We propose that the frequency at f = 0.062 represents biological information for EGF-specific receptor activation, while the frequency at f = 0.288 represents nonspecific, general growth activity.

Once the characteristic frequency of EGF-EGF receptor recognition was identified, it was possible to predict hot spot amino acid positions in the EGF molecule related to this frequency. Using the procedures outlined in Chapter 4, the hot spots in the mEGF were predicted at positions Gly 36, Gly 18, Gly 17, Gly 39 and Gly

5. It is important to note that the predicted hot spots Gly 36, Gly 18 and Gly 39 are conserved throughout the EGF family and the series of EGF-like sequences (Fig. 5.6). These results are correlated here with work done by Campbell and co-workers in 1990 in which they predicted that residues 6, 14, 16, 18, 36, 37, 39 and 42 are important for the structure and function of EGF. When positioned within the 3D EGF structure (highlighted with CPK surfaces and presented in Fig. 5.7), the predicted hot spot amino acid residues appear clustered together in a space around the proposed active site. In addition, the proposed active site, residues 17-46, consists of the B and C loops of EGF. It was experimentally proven that the presence of both the B and C loops is required to activate the EGF receptor: the C loop alone was inactive, while the B loop was weakly active. Furthermore, the RRM predictions are consistent with the finding that only the fragment mEGF-(15-53) was capable of binding to the EGF receptor as well as stimulating DNA synthesis, while the fragment mEGF-(32-53) was less potent in receptor binding activity. It was reported that the fragment TGF-∞-(34-43) was capable of binding to EGF receptor but did not stimulate mitogenic activity. Thus, it is likely that short EGF fragments fail to achieve the active conformation required for receptor binding.

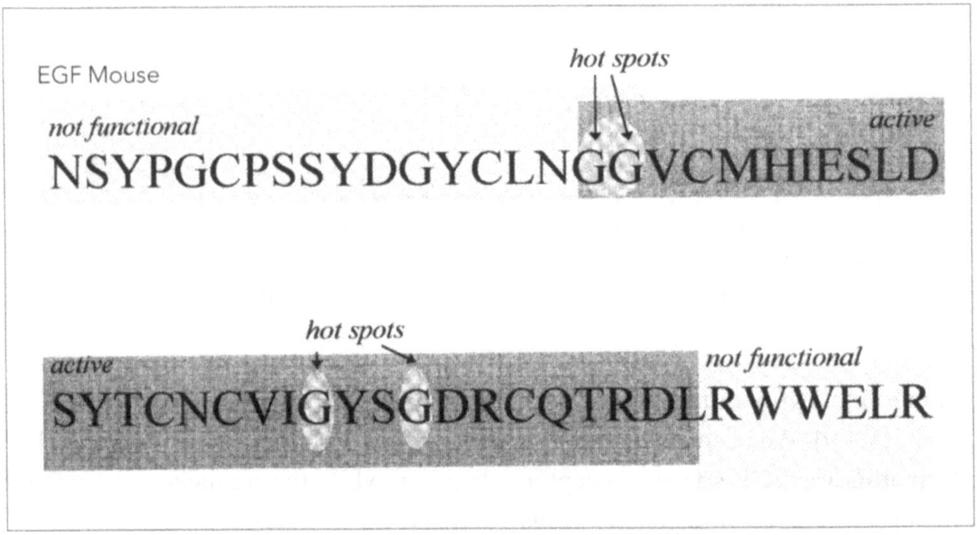

Figure 5.6 Mouse EGf sequence

Light shaded areas (at both terminals) are found to be nonactive. The dark shaded area is proposed as an active site. Circled amino acids are hot spots as predicted by the RRM, and all fall within the proposed active site.

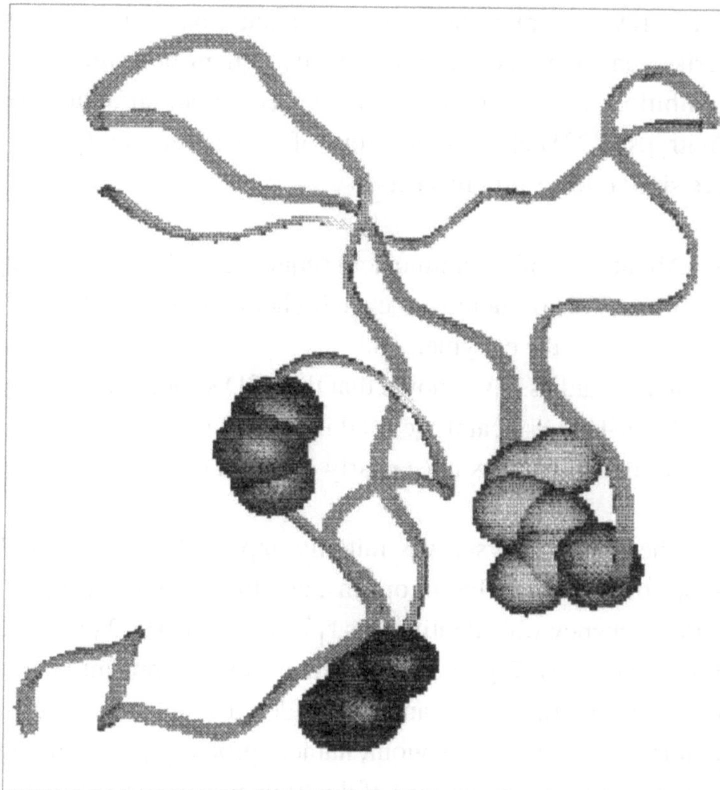

Figure 5.7 Three-dimensional EGF structure with predicted hot spot amino acids highlighted with CPK surfaces
All predicted hot spot amino acids are clustered together in the middle part of the structure, which we propose is an active site of the EGF molecule.

5.4 Active site prediction: example of protease-protease inhibitor interactions

Protein-inhibitor interactions have been studied using the RRM model on an example of a trypsin-trypsin inhibitor complex. Hot spot amino acids, as determined by the RRM, are clustered around the inhibitor binding site and are in close proximity to the enzyme active site. These results are highly relevant for predicting proteolytic and inhibitor binding sites in proteolytic enzymes.

The RRM is applied here to trypsin and chymotrypsin, as they are well-characterised enzymes. This applies equally to their biosynthesis, enzymatic properties, inhibition, kinetics of catalysis and complete structure. Trypsin [72] and chymotrypsin [73] belong to the group of serine proteases (proteolytic enzymes) which are similar in several respects:

- About 40% of the amino acid sequence of these three enzymes is identical. The degree of identity is even higher (about 60%) for residues located in the interior of the enzyme.
- X-ray studies have shown that their 3D structures are very similar.
- A Ser-His-Asp catalytic triad is present in all three.
- All these enzymes have nearly identical catalytic mechanisms [74].

The RRM analysis was initially applied to a group of 18 trypsins and chymotrypsins from different origins, and the common feature in terms of characteristic frequency was identified as $f_1 = 0.3447 \pm 0.002$ (S/N = 258) (Cosic et al., unpublished data) (Fig. 5.8). This frequency component is common to all analysed enzymes and therefore can be considered as the consensus characteristic of their common biological behaviour, namely proteolytic activity, with a catalytic mechanism distinctive in the case of the trypsin and chymotrypsin. This catalytic mechanism is characterised by the Ser-His-Asp catalytic triad among which serine at position 195 is extremely active [74]. In bovine chymotrypsin these three amino acid residues together with catalytic cleft residues from positions 190-226 form an active site [73]. Although catalytic amino acid residues and the active site cleft are not sequentially linked in the 3D protein structure, they are all in close proximity to each other and form a continuous internal cavity.

Having identified the characteristic frequency of trypsins and chymotrypsins, it was possible to further apply the RRM approach to the bovine chymotrypsin sequence to identify amino acids in this sequence which are major contributors to the characteristic frequency. These key amino acid residues in bovine chymotrypsin were located at Gly 184, Gly 187, Gly 193, Gly 196 (the numbering of the amino acid residues corresponds to the Brookhaven Protein database [PDB] entry pdb2cga). All the predicted key amino acid positions are among highly conserved amino acids within chymotrypsin sequences. They are also all sequentially close to or within an active site cleft. Within the bovine chymotrypsin 3D structure [72] all the predicted key amino acid residues are positioned in a cluster in and around

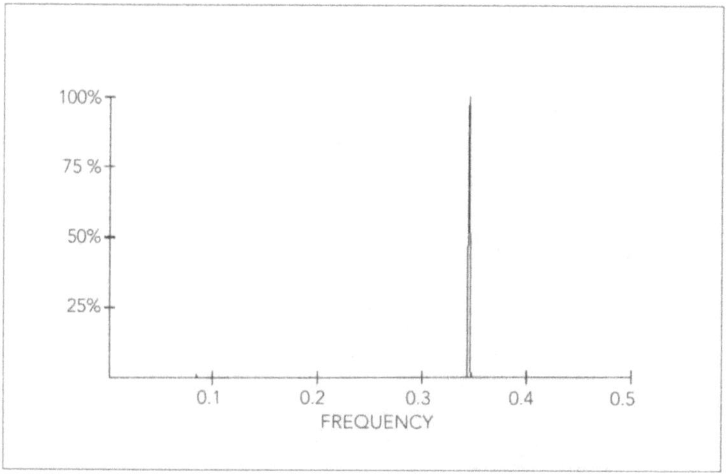

Figure 5.8 Multiple cross-spectral function of 18 trypsins and chymotrypsins from different origins

The abscissa corresponds to the frequency domain of the Fourier transform of the numerical presentation of the proteases, while the ordinate represents the relative intensity of the spectrum. One prominent peak (at frequency 0.3447 ± 0.002 with S/N = 258) in the cross-spectral function denotes only one common frequency component for all the trypsins and chymotrypsins analysed.

the cleft of the protease active site (Fig. 5.9). This result strongly supports earlier conclusions reached with a number of other polypeptides and proteins, that such key amino acid residues predicted by RRM algorithmic procedures are located in spatially clustered regions in the tertiary structure of the polypeptide or protein, and denote a specific biorecognition or interaction site.

The same procedure was repeated with the bovine trypsin. Using the characteristic frequency of trypsins and chymotrypsins, $f_1 = 0.3447$, the key amino acids in the bovine trypsin sequence which are major contributors to the characteristic frequency were identified at Gly 19, Gly 44, Gly 140 and Gly 184 (numbering of amino acid residues corresponds to the Brookhaven PDB entry pdb2tgp). All these amino acid residues are among highly conserved residues within trypsin sequences from different origins. The only exception is crayfish trypsin, which has very little homology with the other trypsins. These amino acid residues are highlighted in the 3D structure of bovine trypsin with CPK surfaces in red. Although they are not sequentially linked, they are clustered in the 3D structure of protease (Fig. 5.10). The

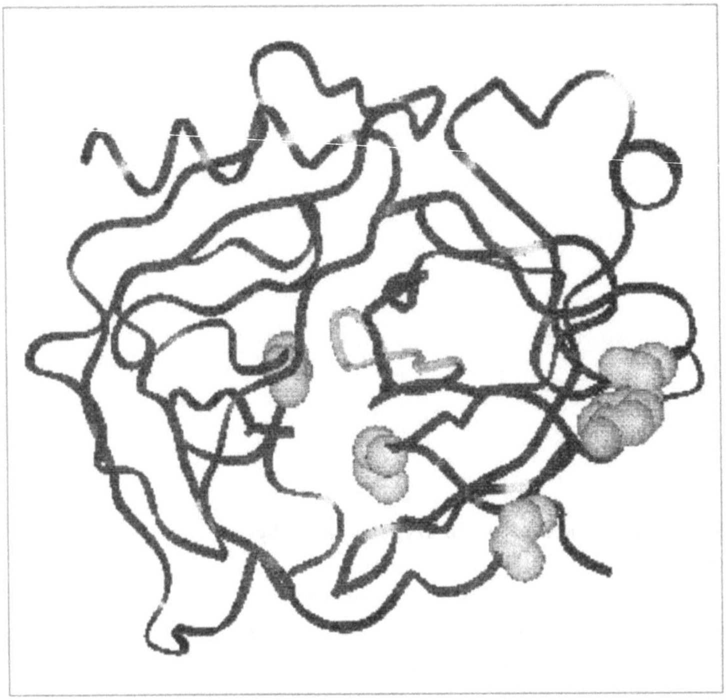

Figure 5.9 Three-dimensional structure of bovine chymotrypsin (PDB entry pdb2cga)

Active site amino acids His 57, Asp 102 and Ser 195 are highlighted with CPK surfaces in blue, while key amino acids predicted by the RRM to be crucial for the specific trypsin/chymotrypsin activity are highlighted with CPKs in yellow. All key amino acids are clustered in the space around the cleft of the active site.

active site of bovine trypsin, which includes amino acid residues Ser 195, Gly 196, Pro 198, Val 199 and Cys 201, is highlighted in light blue, as are the active amino acids His 57, Asp 102 and Ser (Fig. 5.10). As can be seen from Figure 5.10, all predicted key amino acid residues form a cluster on one side of the active site and the active site cleft.

RRM analysis was further performed within the whole group of 80 proteases, including serine (41), cysteine (23), aspartic (47) and metallo (17) proteases, with the aim of identifying a common frequency characteristic which can be associated with the proteolytic function in general. This analysis revealed that the frequency common to all analysed proteases is at $f_2 = 0.3770 \pm 0.002$ with S/N = 510 (Fig. 5.11). Consequently, amino acid residues which mostly contribute to the frequency f_2 were identified in the bovine trypsin sequence at Gly 62, Gly 142, Gly 174 and Gly 193. All these amino acid residues are among conserved residues within trypsins from different origins except Gly 62, which in most cases is Phe 62. When these amino acids are highlighted in the structure of the trypsin-trypsin inhibitor complex, they all appear positioned in a cluster around the inhibitor binding site (Fig. 5.12). In Figure 5.12, the key amino acid positions related to f_2 are highlighted with CPK surfaces in yellow, while trypsin inhibitor is highlighted in light blue. In the same figure, the proteolytically active amino acids are highlighted with CPK surfaces in pink. It is important to note that both the catalytic site and the site where the inhibitor is bound are positioned close to the key amino acids related to the frequency f_2. This observation can be explained in terms of the requirement for structural characteristics related to the general proteolytic function, which can be blocked by an inhibitor (Fig. 5.12).

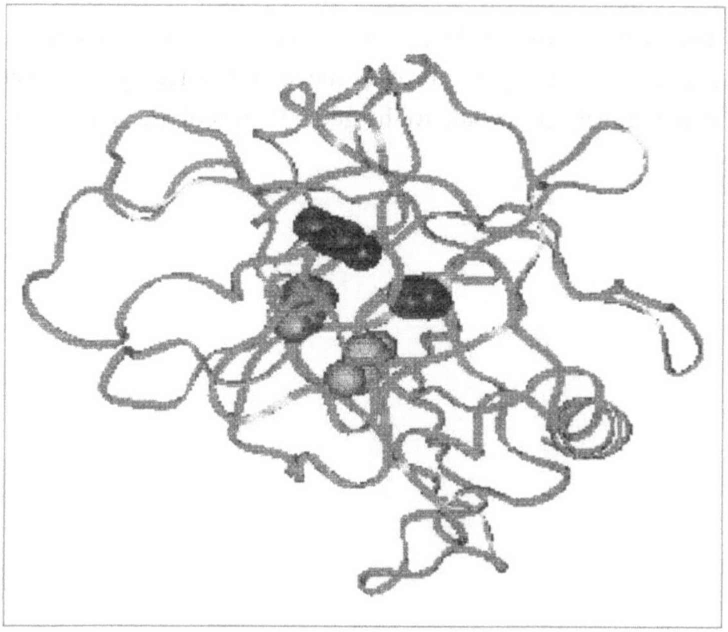

Figure 5.10 Three-dimensional structure of bovine trypsinogen (PDB entry pdb2tgp)
Key hot spot amino acids predicted by the RRM to be crucial for the specific trypsin/chymotrypsin activity are highlighted with CPKs, while the active site is denoted by a lighter ribbon. The active site is positioned close to the cluster of key amino acids.

The RRM analysis of the primary sequences of trypsins and chymotrypsins revealed a frequency characteristic for their specific activity, while the analysis of the whole group of proteases (serine, cysteine, aspartic and metallo proteases) revealed a frequency characteristic for general protease activity. Furthermore, the key amino acid residues which contribute significantly to both characteristic frequencies have been identified within bovine chymotrypsin and trypsin. These amino acid residues are not sequentially contiguous amino acids, but in both cases they are located close to each other in the topographic region over the active site. The prediction of key amino acid residues related to the specific characteristics of trypsins and chymotrypsins for clustering on one side, while key amino acids related to the general protease characteristic cluster on the other side of the catalytic site may infer the necessity of two structural prerequisites for proteolytic activity: one generally associated with proteolysis, and the other associated with the specificity of the particular group of proteases. Protease inhibitors, then, can inhibit proteolytic activity by binding to and possibly structurally modifying one of those important structural sites. This is well illustrated by the case of trypsin and trypsin inhibitor, where the inhibitor is bound to the trypsin just over the spot denoted by the cluster of key amino acids which mostly contribute to the general proteolytic characteristic (Fig. 5.12).

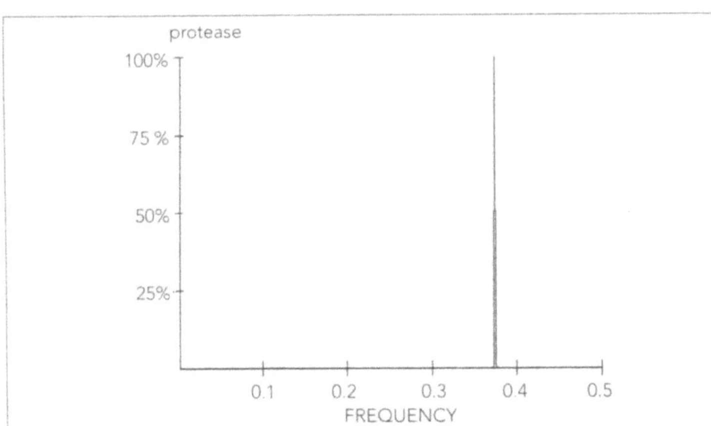

Figure 5.11 Multiple cross-spectral function of 80 different proteases including serine (trypsins, chymotrypsins, elastases, subtilisin and kallikreins), cysteine (papain, actinidin and cathepsins) aspartic (penicillopepsin, acid and renin), and metallo protease from different origins

The abscissa corresponds to the frequency domain of the Fourier transform of the numerical presentation of the proteases, while the ordinate represents the relative intensity of the spectrum. One prominent peak (at frequency 0.3770 ± 0.002 with S/N = 510) in the cross-spectral function denotes only one common frequency component for all the trypsins and chymotrypsins analysed.

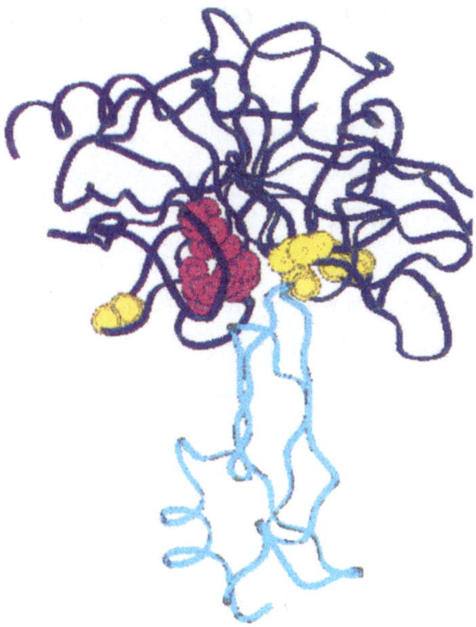

Figure 5.12 Three-dimensional structure of bovine trypsinogen complexed with pancreatic trypsin inhibitor (PDB entry pdb2tgp)

Active site amino acids His 57, Asp 102 and Ser 195 are highlighted with CPK surfaces in pink, while key amino acids predicted by the RRM to be crucial for nonspecific protease activity are highlighted with CPKs in yellow. It can be observed that the active site, as well as the place where the inhibitor is bound to the protease are positioned close to the cluster of key amino acids related to nonspecific protease activity.

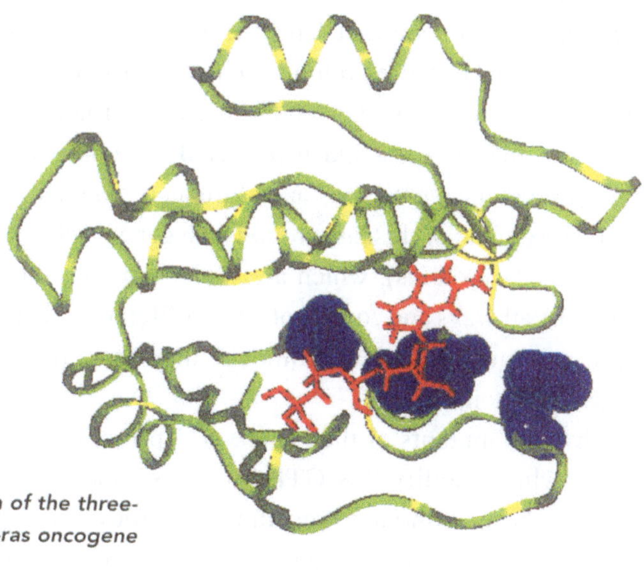

Figure 5.14 Ribbon presentation of the three-dimensional structure of the Ha-ras oncogene product p21

The hot spot amino acids, Gly 10, Gly 12, Gly 13, Gly 15 and Gly 28, are highlighted with CPK surfaces. The active site is denoted by the GTP molecule (in stick presentation) bound to the p21 oncogene product. All predicted hot spot amino acids are clustered around the active site.

5.5 Oncogenes and their activity: example of *ras* proteins

Oncogene proteins are a specific group of growth factors which promote uncontrolled cell growth and proliferation. These proteins are derived from normal cellular growth factors (protooncogenes) via a limited number of modifications (mutations, insertions or deletions). The main question in this section is whether there is a common characteristic of oncogene proteins that characterises their ability to promote uncontrolled cell proliferation. If so, then it would be possible to predict modifications in proto-oncogenes which can transform these proteins into oncogenes. The model protein that is studied here is the Ha-*ras* oncogene product p21.

In recent years, substantial scientific effort has been directed towards unravelling the cellular and molecular mechanisms which lead to the uncontrolled cell proliferation associated with transformation. These investigations have resulted in the identification of a large number of genes (oncogenes), the products of which are involved in cell neoplasia. The oncogenic family of proteins is functionally very diverse. Many of these proteins are found to be related to tyrosine kinases, including the *src*-, *ros*-, *fps*-, *fes*- and *yes*-related proteins, some are DNA binding proteins, e.g. the *myc*-, *fos*- and *myb*-related proteins, whilst the in vivo biological functions for others, e.g. the *ets*- and *sfi*-related proteins still need to be defined [75]. A further example of this last group is the gene product of the Ha-*ras* oncogene family (the p21 proteins), which are GTP-binding proteins involved in signal transduction pathways and control of cell proliferation. The exact cellular role of the p21 proteins remains unknown, although they share fundamental features common to other G-protein systems.

In some mutants of the Ha-*ras* p21 protein structural changes are known to occur which modify this GTP hydrolysis mechanism. Mutations at the active sites of these amino acid residues are believed to underlie the molecular basis of oncogenic transformation [76]. Clearly, procedures which could adequately predict the informational content which is critical for the transforming activity of these oncogenic and proto-oncogenic proteins from their primary sequences would be very useful, not only for further elucidating the structural basis of the function of the proteins per se but also for their application to many other oncogene products. Here, the RRM approach was applied to identify the common characteristic frequency of oncogene products and to relate this characteristic to p21-transforming properties.

The p21 *ras* oncogene products share with a number of other oncogenic proteins an ability to transform cells. The RRM approach was applied to a group of 28 viral and 18 cellular proteins which are products of *myc, myb, mos, fes, fps, fgr, fms, erb, ras, src, abl, yes, syn* and *int*, with the aim of ascertaining their RRM frequency characteristics [32, 33, 64]. As is evident from Figure 5.13, there is one prominent frequency component ($f = 0.0322 \pm 0.004$, $S/N = 468$) in the cross-spectral function, common to the analysed protein sequences related to the Ha-*ras* family. According to RRM axioms, this result suggests that this common frequency characterises a common biological behaviour of this group of oncogene products, ie. their ability to transform cells.

Assuming the intensities of the spectra have a normal distribution, we can calculate the probability that any given frequency amplitude could occur by chance. Using the mean value and the standard deviation for a given spectrum as estimators of a normal distribution, the probability of the cross-spectrum peak at frequency $f = 0.0322$ being due to chance alone is <0.001 (with mean = 0.3105, intensity = 0.7895 Ry, standard deviation = 0.013 and $z = 36.8$).

Once the RRM frequency characteristic had been identified, the hot spot amino acids within the primary sequence of Ha-*ras* p21 were located using the RRM inverse procedure. The hot spot amino acids related to the frequency $f = 0.0313$ were

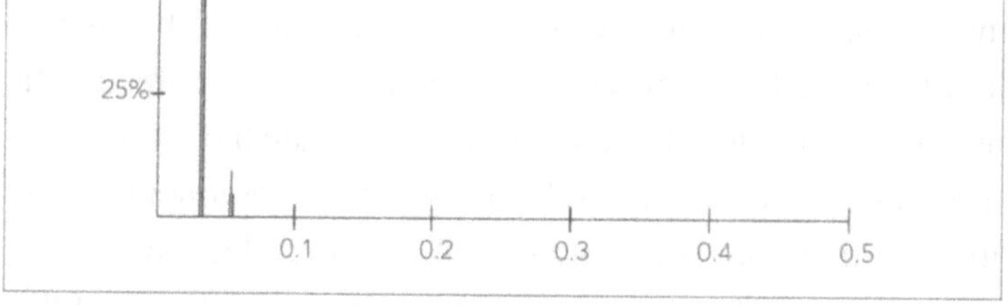

Figure 5.13 Multiple cross-spectral function of 46 different oncogene products

The prominent peak is at the frequency f = 0.0322 ± 0.004.

found Gly 10, Gly 12, Gly 13, Gly 15, Phe 28, Gly 48 and Ser 65. Furthermore, the analysis indicated that the hot spot amino acids at Gly 10, Gly 12, Gly 13, Gly 15 and Phe 28 contribute to the characteristic frequency for p21 oncogenic and GTP-binding function to a greater extent than the amino acids at Gly 48 and Ser 65. In order to validate the proposal that these amino acid residues represent a discrete topological region, the predicted residues were superimposed on the known 3D crystal structure of p21. The results are shown in Figure 5.14 (see page 77).

A comparison between above predictions and other biological and crystallographic findings leads to several significant conclusions. Mutations in the p21 protein at amino acid positions 12, 13 and 61 have been found in a high percentage of human tumors [77]. Furthermore, mutations at these residue positions have a significant effect on the nucleotide dissociation rate constants of p21 and have been implicated in oncogenic activation [76, 77]. Based on the X-ray structure of p21 and its guanine-binding domain [76], it was proposed that the guanine base is sandwiched between the aromatic side chain of Phe 28 and the aliphatic part of the chain of Lys 117. Amino acid residues Gly 13 and Ala 146 were also found to be important for binding the guanine base. As is evident from Figure 5.14, the hot spot amino acids, predicted by the RRM analysis of p21, form a cluster in the tertiary structure, positioned in and around this guanine-binding domain. In particular, it is significant that three out of five predicted hot spot amino acids are found to be among the active site residues (Gly 12, Gly 13, and Phe 28), while the remaining residues Gly 10 and Gly 15, together with Gly 12, Gly 13 and Phe 28, represent part of the continuous topological surface around the guanine-binding site.

Investigations on the physical basis of the RRM indicated that RRM characteristic frequencies represent oscillations of a physical field responsible for energy (information) transfer between interacting molecules. Because the predicted hot spot positions are amino acids in the protein sequence which mostly influence the characteristic frequency, this cluster represents the site where the signal of the characteristic frequency is dominant and thus can act as a resonator for the characteristic frequency. Consequently, this characteristic frequency may dictate the biorecognition specificity of the protein interactions and selectivity of the subsequent energy transfer associated with the functional consequences of the biomolecular interactions. The results with the Ha-*ras* p21 oncogene product provide validation of this concept and a new strategy for characterising and interpreting the information content of oncogenic proteins relevant to cellular transformation.

5.6 Definition of sequence function

The RRM method can be equally applied to proteins, DNA or RNA sequences. However, some problems that arise are dependent on the type of sequence used. For example, proteins are of a defined length, and in most cases their function is distributed along the whole sequence. With DNA, the situation is different. It is a long molecule containing a large number of different genes, including corresponding regulatory sequences; i.e., it contains a large number of functional entities. Their identification within DNA is a problem specific to DNA sequence analysis. On the other hand, while 3D DNA structures can be either helical or extended, proteins can have a whole range of different 3D structures. The 3D structure of protein is determined by the amino acid sequence and is highly relevant to the protein function, as it enables protein active sites to be adequately presented. Thus, a problem specific to proteins is the identification of the active site with regard to the 3D protein structure.

Definition of DNA segment function

Usually, there is a need to identify functional segments within DNA, since it is a long molecule carrying a number of different functional entities (introns, exons, promoters, operators or enhancers). Once the characteristic frequency for a particular function is identified, however, it is possible to identify the corresponding functional fragment within the DNA sequence. This is discussed here with an example of identification of the promoter position within the lacoperon.

A large number of DNA sequences are already known. Aided by current experimental methods and the Human Genome Project, more and more sequences are being identified. However, the critical problem is to identify functional entities within these sequences. Keeping in mind that the length of a DNA sequence can be several thousands of kilobases, it is obvious that the experimental identification of their functional entities would be very laborious and expensive. Thus, a number of predictive, computational methods are being employed to decrease the number of required experiments. One of these methods is identifying the coding sequence by locating start and stop codon positions. Similarly, it is possible to identify the promoter position within a DNA sequence by positioning the promoter core sequence, TATA. However, this characteristic sequence could be found outside the promoter sequence, and thus these predictions are not quite accurate. In general, homology is used to identify other regulatory sequences [13], though the accuracy of these predictions is also often questionable.

In combination with other methods, the RRM has the capability to be a very useful tool in identifying functional entities within the DNA sequence. This is made feasible by scanning the sequence with a window of a certain length and by measuring the frequency spectrum for each position of the window. The window position with the highest peak at the characteristic frequency of the observed function is proposed as the most probable position of the functional entity. The problem is defining the window length. However, most DNA regulatory sequences with the same function have similar lengths. For example, promoters are about 50 nucleotides long.

Identification of the functional entity within the DNA sequence is presented here with an example of the identification of the promoter within the lacoperon sequence:

GCGCAACGCAATTAATGTGAGTTAGCTCACTCATTAGGCACC
CCAGGCTTTACACTTTATTGCTTCCGGCTCGTATGTTGTGTGGAATTGT
GAGCGGATAACAATTTCACACAGGAAACAGCT

The window of 50 nucleotides is used to scan along the sequence. The RRM spectrum is calculated for each position, and amplitude at the promoter characteristic frequency, f = 0.344, is recorded. The window position at which the amplitude at the promoter frequency is the highest is proposed as the most probable promoter position. Table 5.2 presents the amplitudes of the spectra for different segments of lacoperon at the frequency of 0.344.

Data from Table 5.2 are also presented graphically in Figure 5.15. Segments starting from the 40th and 45th nucleotides have the highest amplitude (0.036 Ry^2) at the characteristic frequency, and these segments are predicted as possible promoter sites. The second predicted promoter position is within two nucleotides of the one determined experimentally. Figure 5.16 represents spectra of analysed segments along the lacoperon sequence. The segment with the highest amplitude at the promoter characteristic frequency is marked with an arrow. Most of these spectra

***Table 5.2* Amplitude of the spectrum for each 50-nucleotide segment along the lacoperon sequence**

The position refers to the starting position of each segment.

Position	Amplitude Ry^2	Position	Amplitude Ry^2
0	0.0095	30	0.015
5	0.018	33	0.019
11	0.005	35	0.023
12	0.005	38	0.027
13	0.003	40	0.036
15	0.004	43	0.032
18	0.000	45	0.036
20	0.009	48	0.012
23	0.008	50	0.009
25	0.016	53	0.02
28	0.014	55	0.027

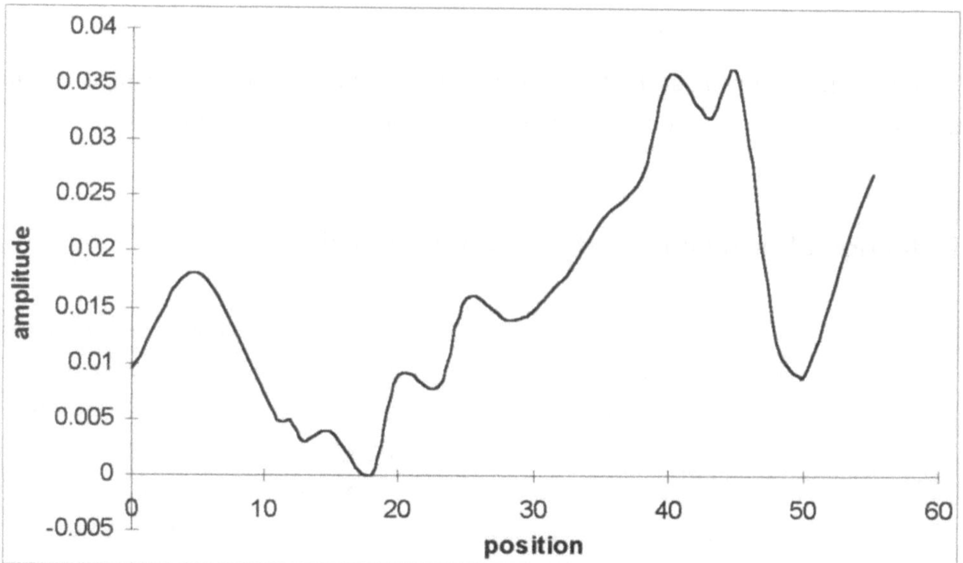

***Figure 5.15* Graphical presentation of amplitudes of the spectra of 50-nucleotide windows at the promoter characteristic frequency versus starting positions of the windows**

The maximum is reached with windows starting from the 40-45th nucleotide, which is approximately the position of the promoter.

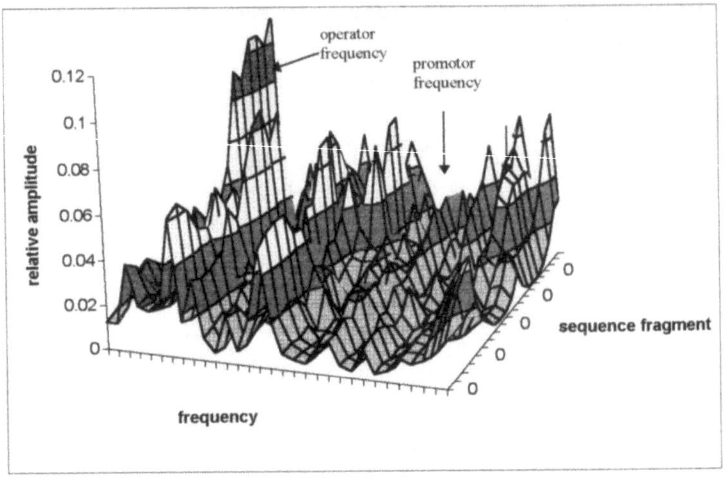

Figure 5.16 Spectrum of each analysed segment along the lacoperon sequence

The segment with the highest amplitude at the promoter characteristic frequency is marked with an arrow. The operator frequency is also marked with an arrow.

have their highest peak at the frequency characteristic for operators. This is not at all unexpected, since lacoperon contains operators as well as promoters.

Definition of protein or protein segment function

The procedures described above in the example of DNA could be applied to the definition of protein function or functionally important fragments of proteins. The identification of the whole protein function is presented here with the tumor necrosis factor as an example. Functional fragment identification is presented, with an example of HIV envelope protein. Identification of fragments responsible for binding to CD4 antigen and the possible prevention of these and other deteriorating functions of HIV gp120 protein are also discussed.

A group of proteins, TNFs, was chosen as a model system to illustrate the proposed methods of identifying protein function [24, 35]. These proteins represent a defined group of extracellular factors with multipotent activities, some important for dis-

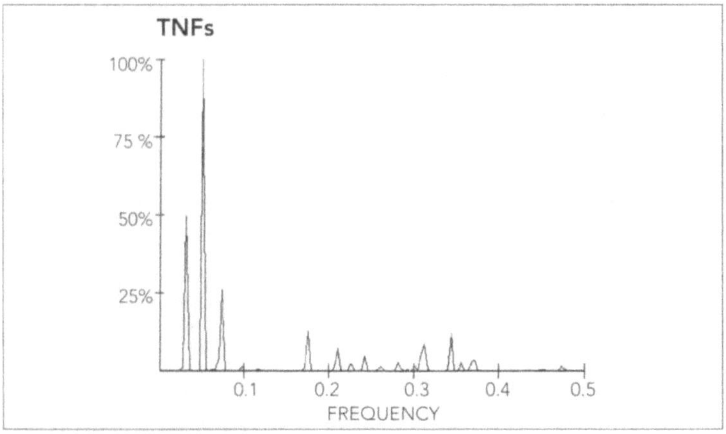

Figure 5.17 Multiple cross-spectral function of eight TNFs from different origins: human, bovine, mouse and rabbit
Three prominent frequencies occur.

ease treatment: necrosis of tumor tissue. The sequential evaluation of the consensus spectrum of TNFs is presented in Figure 5.17. Two common frequencies which mostly characterise their multipotent functions can be seen. The consensus spectrum of TNFs was then compared with the characteristic frequencies already defined. The result is presented in Figure 5.18. One can see that these two frequencies are the same as the characteristic frequencies for oncogenes (transforming proteins) and for DNA regulation via enhancers. These results are in accordance with the fact that TNFs can regulate the production of other proteins (possibly by DNA regulation) and also mediate necrosis of transformed tissue.

In 1992 Veljkovic and co-workers 1992 [78] used a similar approach to identify similarities between vasoactive intestinal peptide (VIP) and the second conserved region of the human immunodefficiency virus type 1 (HIV-1) envelope protein gp120. They have investigated the following conserved regions within the gp120: RSANFTDNAKTIIVQLNESVEINCTRP (peptide pC2); its segment RSANFTDNAKTIIVQLNESVEIN (peptide NTM); and the central portion of the peptide NAKTIIVL (peptide V). Comparing the RRM spectra of gp120 HIV proteins with the spectrum of peptide V, these authors have identified peptide V as a potential CD4-binding domain [78].

In addition, the same authors have investigated the peptide ASTTTNYT (peptide T) which was found in the SF2 isolate of ARV-2. When RRM spectra were calculated for these peptides, it was shown that peptide NTM, peptide T and VIP (va-

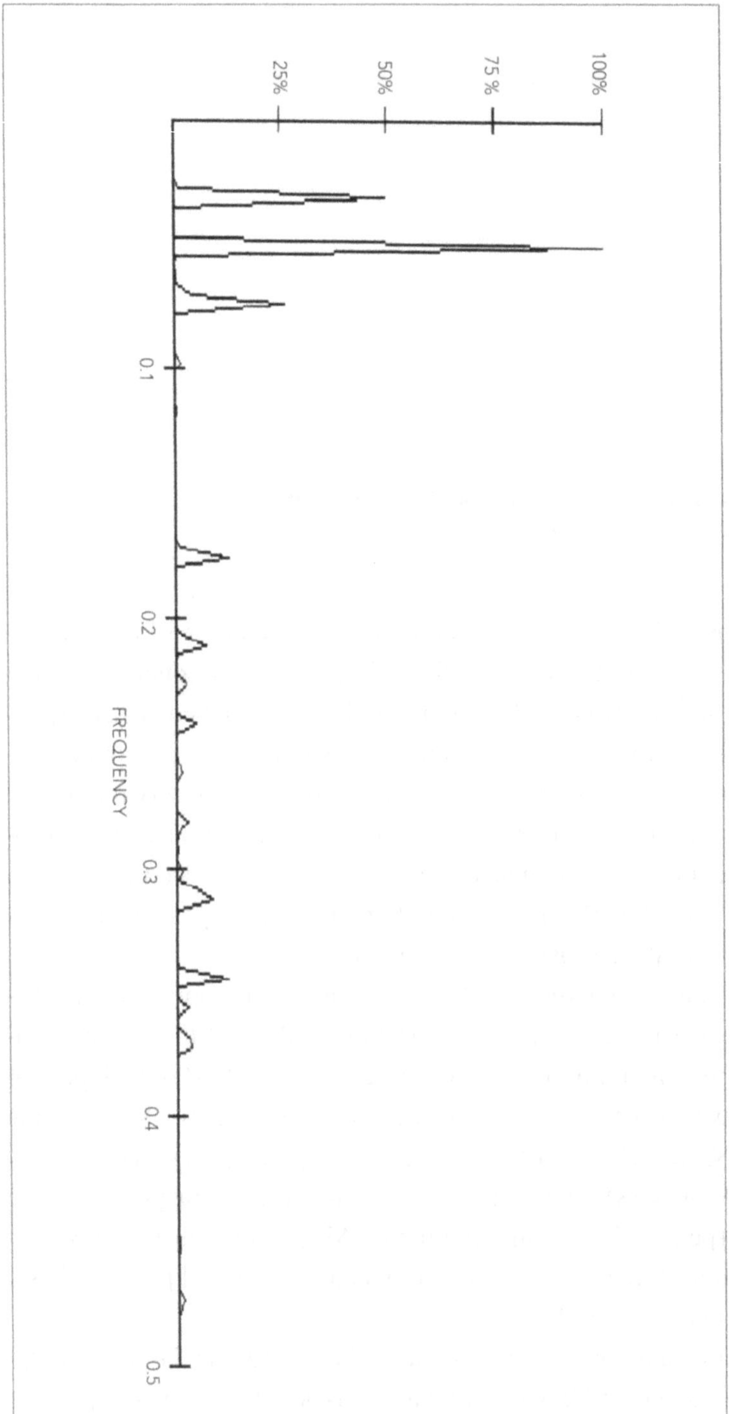

hemoglobins

oncogens
protein A
phospholipases
homeo box proteins
DNA regulation via enh.

cytochromes B
neurotoxins
myoglobins
interferons

bacterial repressors
heat shock proteins
repressors
phage repressors

HIV gp 160

HIV p17, gp 120

competence growth factors

restriction enzymes
growth factors

glucagons
lysozymes

RNA polymerase
protease inhibitors
proteases

NGFs
amylases
kinases
fibrinogens
FGFs
Cytochrome C

ACH receptors
IGFs

sointensive peptide) do have striking similarities in their RRM spectra. With regard to these results, authors have speculated that "(i) VIP could serve as a possible antigen eliciting antibodies recognising the peptide pC2. (ii) the antibodies recognising peptide pC2 could block viral fusion and (iii) the peptide NTM and its derivatives could be applied to halt or attenuate the spread of virus in infected individuals, as well as to the development of AIDS vaccine components."[78]. These speculations are in accordance with findings that (1) VIP prevents the neuron-depleting action of HIV-1 gp120, (2) peptide T blocks the adsorption of the viron to CD4 receptor in human peripheral blood T4 lymphocytes and (3) antiserum to peptide T prevents the neuronal cell death produced by HIV-1 gp120.

◀ ***Figure 5.18 Multiple cross-spectral function of TNFs as defined in Figure 5.17 compared with all characteristic frequencies already identified***

Three prominent frequencies are representing oncogene, enhancer (DNA binding) and toxins characteristic frequencies.

6 Protein Interactions

Protein function is achieved through interaction of a protein with its target(s). Usually these targets are other proteins: receptors, activators, inhibitors or antibodies. Should these interactions be based on resonant recognition between interacting proteins, as is proposed with the RRM, then we can expect proteins and their targets to have a common characteristic frequency. This is discussed here, and a number of examples are presented. Phases at recognition frequency are also analysed.

6.1 Protein-protein interactions: frequency requirements

The main postulate of the RRM is that protein-protein functional recognition and interaction is resonant at the frequency characteristic for both the protein and its target. Examples of protease-protease inhibitors and insulin-insulin binding proteins are presented.

Interaction of protease-protease inhibitors

When RRM analysis was performed on an entire group of 80 proteases (all of them having proteolitic function; i.e., they can selectively digest other proteins), the common frequency was identified at $f_1 = 0.377 \pm 0.005$ with S/N = 510 (Fig. 6.1a). In general this frequency is associated with proteolitic function (protease inhibitors are proteins that inhibit protease proteolitic function). If proteolitic function is characterised with the frequency $f_1 = 0.377 \pm 0.005$, it is reasonable to expect that the proteolitic function can be inhibited by interference at this frequency. Thus, 27 different protease inhibitors were analysed using the RRM to find out if their func-

 ©1997, Birkhäuser Verlag Basel

tion had a characteristic frequency, and if this frequency corresponded to the proteolitic characteristic frequency. The characteristic frequency of protease inhibitors was found to be at $f_2 = 0.355 \pm 0.017$ with S/N = 203 (Fig. 6.1b). The results presented show that characteristic frequencies of proteases and protease inhibitors are very close to each other and that they overlap with each other within calculation error. This indicates that inhibition of protease by interaction between proteases and their inhibitors is based on the interference by inhibitors of the protease activity at the same frequency.

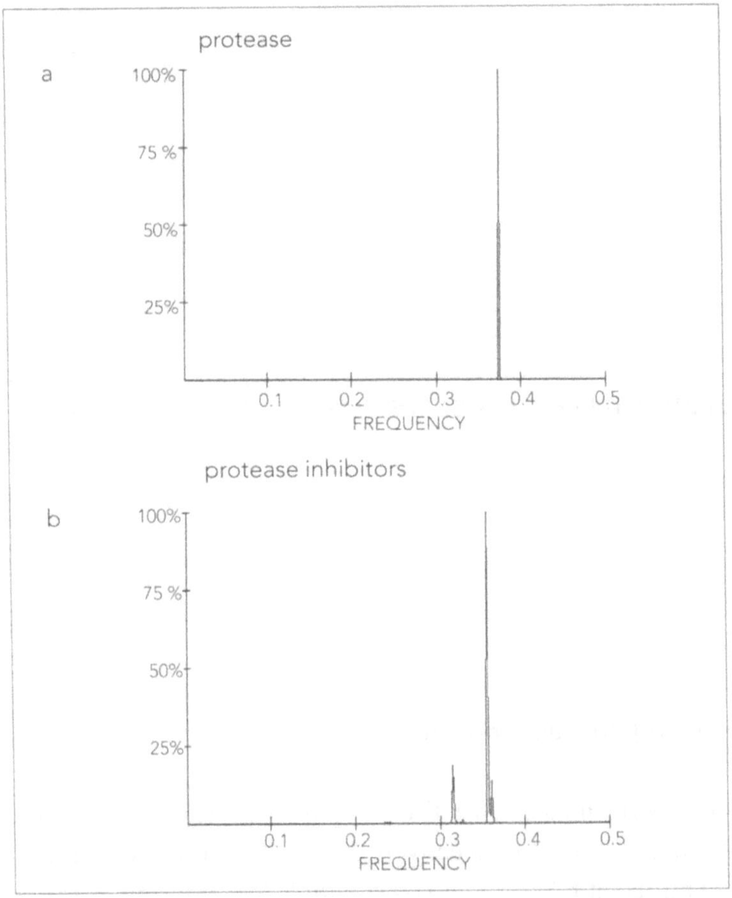

Figure 6.1 Comparison of the RRM characteristic frequencies of interacting groups of proteins

Interaction of protease-protease inhibitors: (a) Consensus RRM spectrum of 80 different proteases. Common characteristic frequency is $f_1 = 0.377 \pm 0.005$ with S/N = 510. (b) Consensus RRM spectrum of 27 different proteases. Common characteristic frequency is $f_1 = 0.355 \pm 0.017$ with S/N = 203.

Interaction of IGF-IGF binding proteins

The IGFs have several important functions *in vivo*, e.g. stimulating cell growth and differentiation. In circulation, most IGFs are bound and regulated by six different IGF-binding proteins, IGFBPs [79]. The IGFBPs vary in their affinities for the different IGFs, with IGFBP-2, 5 and 6 preferentially binding IGF-II. The actions of the IGFBPs are generally inhibitory, although IGFBPs may stimulate IGF action, presumably by binding to the cell surface or extracellular matrix and increasing local IGF concentrations. Both heparin and integrin binding motives are found in some of the IGFBPs. Well-characterised mutant forms of IGFs as well as NMR data have provided insight into residues in the IGFs that are involved in binding to receptors and to IGFBPs. However, the 3D structure of IGFBPs has not been solved, and no specific residue in any IGFBP has yet been shown to be involved in interaction with an IGF. It is anticipated, however, that parts of the conserved regions in the N- and C-terminal domains of the molecules constitute the IGF-binding site.

Characterising the IGF system requires a detailed understanding of the molecular mechanisms involved in IGF/IGFBP interactions. In order to determine structurally and functionally important residues and regions in IGFBPs, several strategies are available, such as site-directed mutagenesis of conserved residues. However, there is approximately 30-40% homology between different IGFBPs, and because of their size, the more information available to effectively perform those studies, the better.

Six human IGF-binding proteins (IGFBP-1, -2, -3, -4, -5 and -6) have been analysed using the RRM (Cosic and Forsberg, unpubl. data). The multiple cross-spectrum of these six proteins revealed one prominent peak at the frequency $f_1 = 0.1624 \pm 0.001$ (Fig. 6.2a.). This frequency component is common to all proteins analysed, and thus we propose that it be related to their common biological function, i.e. IGF binding.

The frequency that characterises IGFBP is different from the one that characterises IGFs. The characteristic frequency obtained previously for IGFs was $f_2 = 0.492 \pm 0.008$ [29] (Fig. 6.2b). Different characteristic frequencies for IGFs and IGFBPs are expected to occur because these proteins have different biological functions. On the other hand, the characteristic frequency of IGFs is related to their main common biological function, growth promotion and recognition of corresponding receptors, while the characteristic frequency of IGFBPs is related to their common ability to bind IGFs.

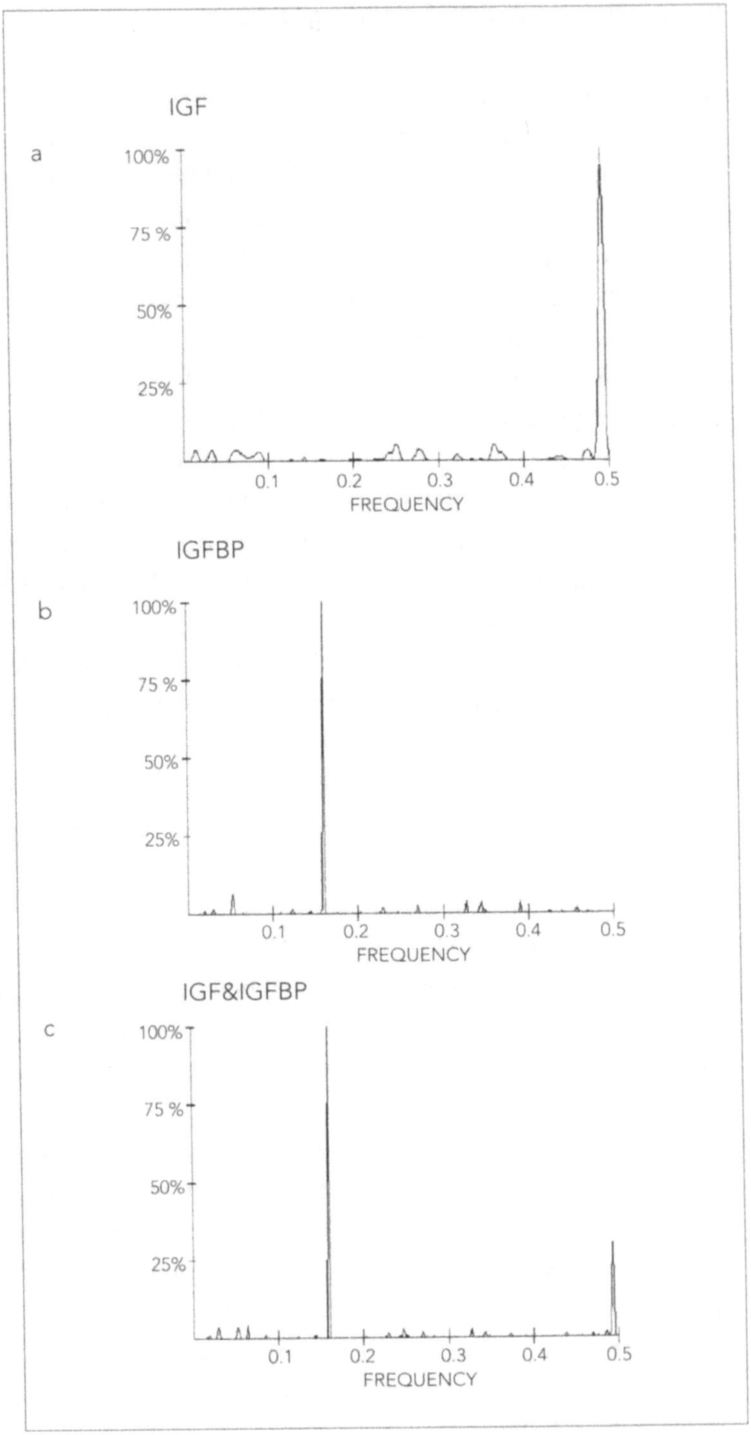
IGF
a
100%
75 %
50%
25%
0.1
0.2
0.3
0.4
0.5
FREQUENCY
IGFBP
b
100%
75 %
50%
25%
0.1
0.2
0.3
0.4
0.5
FREQUENCY
IGF&IGFBP
c
100%
75 %
50%
25%
0.1
0.2
0.3
0.4
0.5
FREQUENCY

Figure 6.2

To confirm that IGFBP frequency is related to IGFBP binding to IGF, we compared IGFBPs and IGFs using a multiple cross-spectral function as defined by the RRM. The comparison revealed that the frequency $f_1 = 0.1624 \pm 0.001$ is the one most common to IGFBPs and IGFs (Fig. 6.2c) and is most likely characteristic of their mutual binding. However, the other frequency, $f_2 = 0.492 \pm 0.008$, was also found to be common to IGFBPs and IGFs, though less prominent. Thus, theoretically f_2 could also be considered an important characteristic for neutralisation of IGF activity within the IGF-IGFBP complex. This result is similar to the result obtained previously with proteases and protease inhibitors. However, in this case the inhibitor frequency was found to be different from the main functional frequency. This can be explained by the different functions of inhibitors and binding proteins: inhibitors inhibit the function, while binding proteins only moderate (suppress) the function (Cosic and Forsberg, unpublished data).

6.2 Protein-protein interactions: phase requirements

We have established that interacting proteins recognise each other on the basis of the same characteristic frequency. However, the question arises, Why do proteins that recognise the same target and have the same characteristic frequency not recognise and interact among themselves? An additional question would be: If a functional group of proteins and their target proteins do have the same characteristic frequency, would it mean that every protein from this group can interact with every protein from the target group? How would species specificity ever happen? Answering these questions requires an additional criterion for mutual interaction between proteins. This parameter has been found in the phase spectra of numeric representations of proteins. As discussed previously, it was found that the phases at the characteristic frequency are approximately opposite for each pair of interacting proteins. We discuss this below and present examples.

◀ ***Figure 6.2 Comparison of IGFs and their binding proteins as two interacting families of proteins***

(a) Consensus RRM spectrum of different IGFs. Common characteristic frequency is at $f_2 = 0.492 \pm 0.008$. (b) Consensus RRM spectrum of six different IGFBPs. Common characteristic frequency is at $f_1 = 0.1624 \pm 0.001$. (c) Cross-spectral function of IGFs and IGFBPs. Two common frequencies are revealed at $f_2 = 0.492 \pm 0.008$ and $f_1 = 0.1624 \pm 0.001$, where f_1 is more prominent.

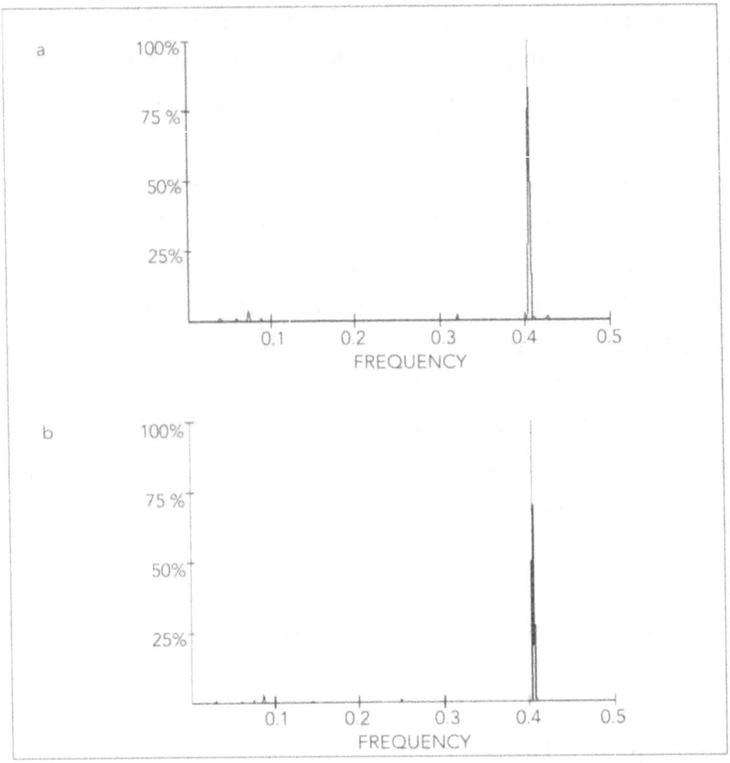

Figure 6.3 Comparison between NGFs and their receptors
(a) Consensus RRM spectrum of eight NGFs. Common characteristic frequency is at $f_1 = 0.404 \pm 0.008$. *(b) Cross-spectrum of eight NGFs and seven NGF receptors. Common frequency is again at* $f_1 = 0.404 \pm 0.008$*, denoting the characteristic for their mutual interaction.*

Interaction of NGF-NGF receptors

Nerve growth factor is essential to the maintenance and differentiation of sensory and sympathetic neurones. Receptors for NGF are present in a variety of normal sympathetic and sensory neurones as well as in normal and neoplastic chromaffin cells.

RRM analysis was applied to a group of eight NGFs including snake venom and mammalian factors. The analysis revealed a prominent common frequency characteristic at $f = 0.404 \pm 0.008$ with S/N = 192 (Fig. 6.3a). To determine whether this frequency is characteristic of NGF-receptor recognition, NGFs were compared with their receptors, and an even more prominent peak (S/N = 282) occurred at the same frequency (Fig. 6.3b). Phases at the characteristic frequency were found to be $\varphi = -0.34$ rad and $\varphi = 2.62$ rad for human NGF and its receptor, respectively. The phase difference was $\Delta\phi = 3.96$ rad, close enough to π, a result which is in accordance with RRM postulates that proteins and their receptors have opposite phases at the characteristic frequency [26].

Interaction of IL-2–IL-2 receptors

The next example of protein-protein receptor interaction is illustrated with the example of IL-2 and its receptors. The multiple cross-spectral function of IL-2 sequences and corresponding receptors (human and mouse) revealed one common characteristic frequency at $f = 0.308 \pm 0.008$. When phase spectra between interacting protein and protein receptor were observed, phase differences at the characteristic frequency were found to be 2.77 rad between human IL-2 and its receptor and 1.45 between rat IL-2 and its receptor. These values are close enough to π, i.e. phases for IL-2s and their receptors are almost opposite [24].

Interaction of immunoglobulin V_{HIII} and protein A

Immunoglobulins are among the most complicated proteins. Their natural function is to bind very selectively. Thus, immunoglobulins are used in a number of technological processes, such as affinity binding, where their ability to bind very selectively to other molecules is used in the separation of particular molecules from the rest of a solution. Understanding the relationships among the sequence, structure and selectivity of the antibody is one of the most difficult problems in structural biology. Once these relationships are understood, it would be possible to modify immune reactions within a body as well as to improve and rationalise affinity binding processes.

V_{HIII} regions of immunoglobulins are regions responsible for selective binding to the antibody. The V_{HIII} region that binds to protein A is chosen as a model example for the analysis of selectivity of V_{HIII} regions interactions because some V_{HIII} regions are protein A binding and some are nonbinding, although all of these regions are very homologous [80]. Until now, there has been no effective way of predicting separation of binders from nonbinders. To find out if there is any difference between protein A-binding and nonbinding V_{HIII} regions, the RRM was applied to binders and nonbinders separately. The multiple cross-spectral function of human binders (seven sequences from ref. 80) revealed only one prominent peak at the frequency $f_1 = 0.027 \pm 0.01$ (Fig. 6.4a.). The frequency of this peak thus ranges from 0.017 to 0.037. When nonbinders (four sequences from Runden, ref. 80) were compared, a few prominent peaks occurred. The most prominent peak occurred at the frequency $f = 0.45 \pm 0.01$, and a less prominent one occurred at the frequency $f_1 = 0.027 \pm 0.01$ (Fig. 6.4b), implying that nonbinders have characteristics different from binders (Cosic and Korsnes, unpublished data). To determine whether one or the other frequency characteristic was related to the binding to protein A, proteins A were compared among themselves. Only two protein A variants have been found in the database, and when the RRM cross-spectral analysis was applied to these two proteins, the resulting cross-spectrum was still noisy. However, there is one most prominent peak at $f_2 = 0.035 \pm 0.002$ (with peak frequency range from 0.033 to 0.037) (Fig. 6.4c), and this frequency can be considered to be one of the protein A characteristics. This characteristic for is within the same frequency range (within calculation error) as the characteristic of V_{HIII} that binds the protein. In addition, this characteristic is not the most prominent one among nonbinders. Thus, keeping in mind the RRM postulate that specific binding is always characterised by the frequency common to both participants in the binding (interaction), we propose that frequency $f_1 = 0.027 \pm 0.01$ is characteristic of the binding of protein A to V_{HIII} regions (Cosic and Korsnes, unpublished data).

To check whether this frequency is indeed characteristic of protein A binding to V_{HIII} regions, mouse V_{HIII} binders and nonbinders were compared in the same manner as human ones. As can be seen in Figure 6.5a and b, the situation with mouse binders and nonbinders is even clearer than with human ones. The cross-spectral function of 12 mouse binders revealed only one common peak at the same frequency as for human binders: $f_1 = 0.027 \pm 0.01$. However, when nonbinders (14 sequences from ref. 81) were compared, a completely different common frequency was obtained. The frequency $f_1 = 0.027 \pm 0.01$ was not observed among the first

Figure 6.4 Comparison between protein A-binding and nonbinding human VHIII regions ▶

(a) Cross-spectral function of seven VHIII regions that bind to protein A. Peak frequency is at f1 = 0.027 ± 0.01

(b) Cross-spectral function of four V_{HIII} regions that do not bind to protein A. Peak frequency is at $f_2 = 0.450 \pm 0.01$. (c) Cross-spectral function of two proteins A. Peak frequency is at $f = 0.027 \pm 0.01$, indicating that this frequency is crucial for binding between protein A and V_{HIII} regions.

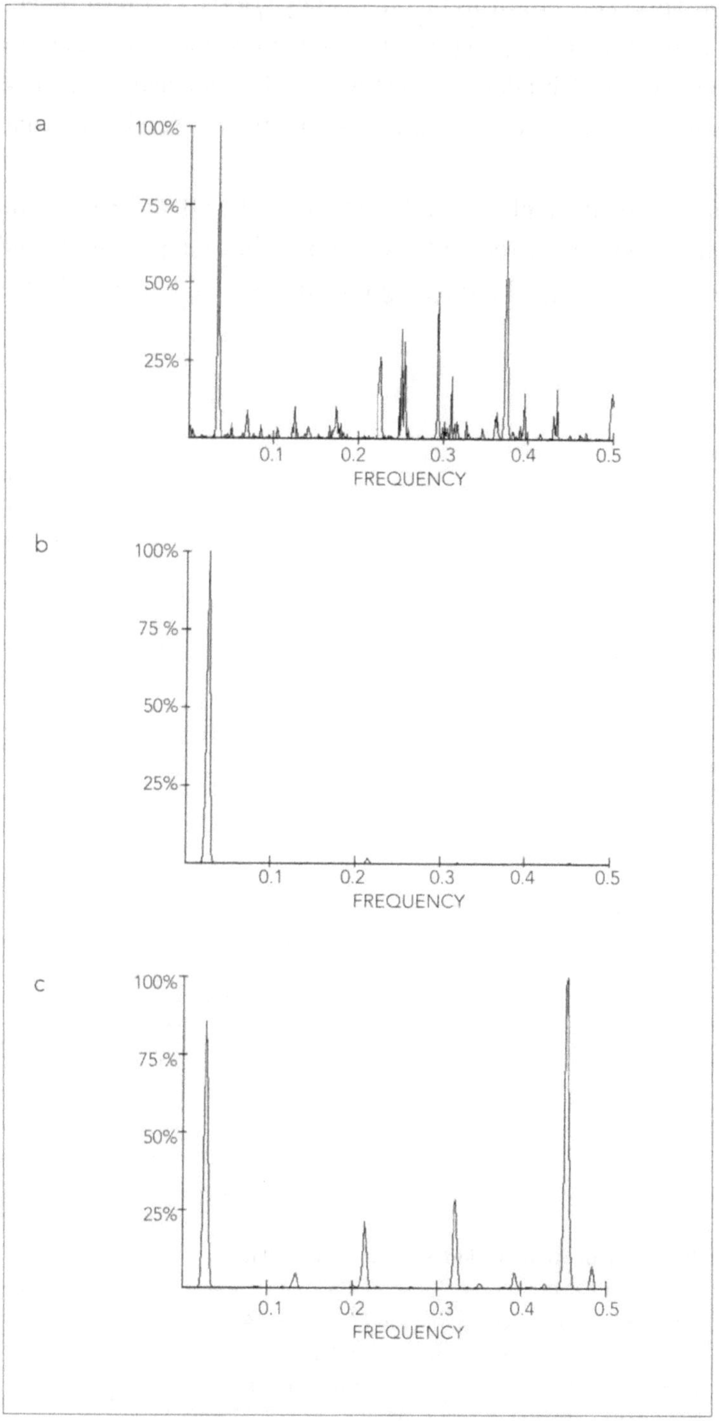
a
100%
75 %
50%
25%
0.1
0.2
0.3
0.4
0.5
FREQUENCY
b
100%
75 %
50%
25%
0.1
0.2
0.3
0.4
0.5
FREQUENCY
c
100%
75 %
50%
25%
0.1
0.2
0.3
0.4
0.5
FREQUENCY

three prominent peaks. This result confirms that frequency $f_1 = 0.027 \pm 0.01$ indeed characterises protein A binding to V_{HIII} regions. Consequently, these results show that distinction between protein A binders and nonbinders detected using the RRM method is simple despite a high degree of mutual homology between binders and nonbinders.

The next step was to investigate phases at the frequency $f_1 = 0.027 \pm 0.01$ and to test whether they are opposite between interacting proteins, i.e. protein A- and V_{HIII}-binding regions. For all 17 V_{HIII}-binding regions (mouse and human), the av-

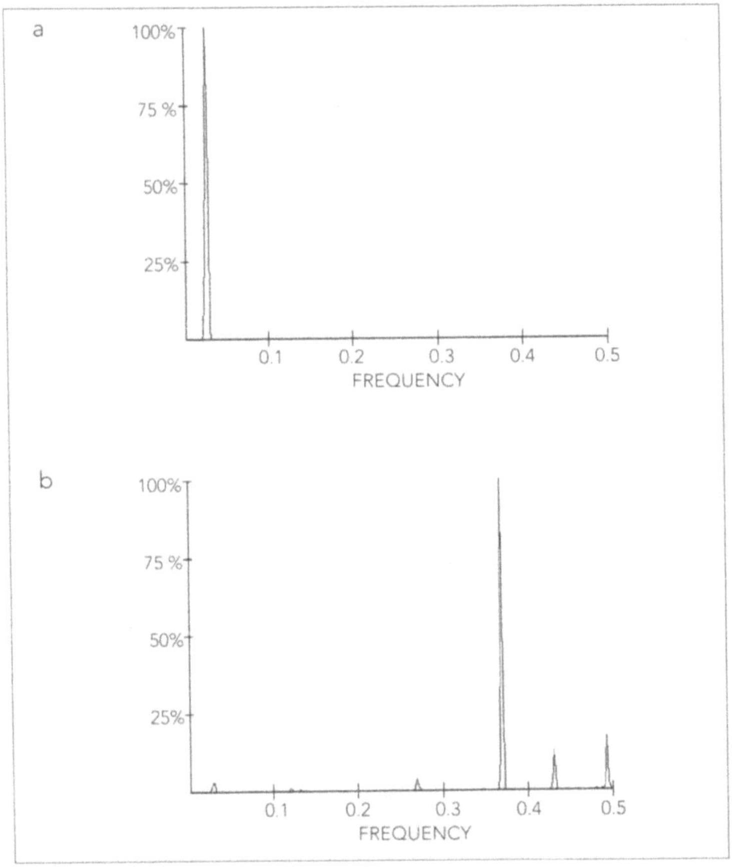

Figure 6.5 Comparison between protein A-binding and nonbinding mouse VHIII regions

(a) Cross-spectral function of 12 mouse V_{HIII} regions that bind to protein A. Peak frequency is at $f_1 = 0.027 \pm 0.01$, the same obtained with human sequences. (b) Cross-spectral function of 14 V_{HIII} regions that do not bind to protein A revealed quite different peak frequencies.

erage phase at $f_1 = 0.027 \pm 0.01$ was $\varphi ave1 = 1.65$ rad with standard deviation (SD) = 0.35. However, there is a large variation in the phase at the same frequency for nonbinding V_{HIII} regions, where average phase value was found to be $\varphi ave2 = 1.06$ with SD = 1.46. As the SD is greater than the average value, this average value is not significant. On the other hand, the phase at $f_1 = 0.027 \pm 0.01$ in the first protein A (A1) sequence was $\varphi_{a1} = -0.725$ rad, while in the second protein A (A2), it was $\varphi_{a2} = 1.437$ rad. The phase differences between each protein A sequence and the average phase of binding V_{HIII} regions were $\Delta\varphi_{a1} = 2.34$ rad and $\Delta\varphi_{a1} = 3.047$ rad, both of which are reasonably close to π. This means that the phase at the frequency that characterises protein A binding to V_{HIII} regions is opposite in protein A to that in V_{HIII} regions. This is one among a number of examples where interacting proteins have the same characteristic frequency with opposite phases between interacting pairs at that frequency.

6.3 Protein-DNA interactions

The regulation of gene expression plays a central role in the cellular processes of proliferation, differentiation and transformation. Specific protein-DNA interactions are known to play a critical role in gene regulation. Investigations in this field have largely been directed towards an understanding of these interactions through the identification of structural features of proteins and DNA segments involved in these processes. However, these model investigations were not able to anticipate the physico-chemical parameters which characterise the molecular forces responsible for selective recognition between proteins and particular DNA segments.

The structural features of protein-DNA interactions have been evaluated using the RRM in conjunction with molecular graphics techniques. The physicochemical basis of this analysis was first validated with the trp repressor-operator interaction as a well-defined example. The amino acid and the structural features predicted by these procedures to be crucial to the repressor-operator interaction were found to be clustered around the known 3D structure of the active site of the trp repressor. Similar methods of analysis have been extended to the less well defined example of the Ha-*ras* p21 protein family. The results of this analysis have indicated two distinct interactive regions in p21: one associated with the guanine binding site and a second which is proposed to be associated with a binding site for an activator protein. These studies indicate that the p21 protein, in addition to its ability to function as a plasma membrane-associated G-protein and to bind free guanine nucleotides in cytoplasm, has the structural ability to bind guanine incorporated in DNA. Thus, p21-related proteins may have the potential to function as DNA-binding and regulating proteins with the mode of upstream DNA binding closely related to their oncogenic function.

In recent years, substantial scientific efforts have been directed towards unravelling the cellular and molecular mechanisms which lead to uncontrolled proliferation associated with cell transformation. As a generally accepted model, selective protein-DNA recognition is often discussed in terms of mutual structural complementarity, with the protein and DNA fitting together like a lock and key. Although this model has been used [82, 83] with a number of examples of protein-DNA interactions, the "lock-and-key" model cannot predict the critical shapes which allow the interaction, identify *de novo* the amino acids or nucleotides which are crucial in defining this shape, or anticipate the overall physicochemical parameters which characterise the molecular forces responsible for selective recognition between proteins and certain sequence arrangements of DNA. Thus, the lock-and-key model has important empirical but no predictive capabilities relevant to the submolecular basis of protein-DNA recognition.

To gain insight into the physico-chemical and structural requirements which may characterise protein-DNA interactions, the usage of the RRM was evaluated. The investigations of the characteristic RRM frequencies and their biological significance have demonstrated that proteins and their protein or DNA binding targets exhibit the same characteristic frequency [24, 27]. Thus, it can be proposed that this characteristic frequency is associated with recognition between interacting biomacromolecules. The mechanism of protein-protein or protein-DNA interaction can be comprehensively described in terms of a complex system of electromagnetic oscillators. The requirements for specific protein-target complex formation can thus be considered to involve the coupling of the electromagnetic oscillations of the protein and its target. As discussed in Chapter 3, the RRM characteristic frequencies represent oscillations of the physical field responsible for specific recognition and resonant energy transfer between interacting biomacromolecules.

Once the characteristic frequency corresponding to a specific functional property of a protein family is defined, it is possible to utilise the RRM approach to predict amino acids in the sequence which predominantly contribute to this frequency. These predicted amino acids are thus likely to be crucial to the function and to protein activity in a number of examples including IL-2 [29] and the SV40 enhancer [28]. In studies with hemoglobins, myoglobins, lysozymes [31], transforming proteins [32] and a number of other protein families including cytochrome C [25], IGFs and glucagons [25], it has been shown that in each molecule predicted hot spot amino acids form a unique cluster positioned around a functional site in the tertiary structure of the protein.

Protein-DNA interactions: structural requirements

As a generally accepted model, selective protein-DNA recognition is often discussed in terms of mutual structural complementarity, with the protein and DNA fitting together like a lock and key. Although this model has been used prospectively with a number of examples of protein-DNA interactions, the "lock-and-key" model cannot predict the critical shapes which allow the interaction.

The present study extends the RRM approach to the analysis of the structure and function relationships of DNA regulating proteins using the trp repressor as an independently well documented example. The trp repressor is known to regulate the synthesis of L-tryptophan by binding to the corresponding operons [84]. Repression of these operons is regulated by a negative feedback loop: as the concentration of L-tryptophan increases, two molecules of L-tryptophan bind to the inactive trp aporepressor, a dimer of two identical 107-*mer* subunits, forming an active trp repressor. The spatial disposition of the predicted hot spot amino acids in the trp repressor associated with the operator binding sites was analysed using a combination of RRM methods and molecular graphics procedures. Furthermore, topographic regions of the trp repressor, which are predicted to be relevant to DNA interaction, have been compared with the 3D structure of the bound DNA segment [65, 84]. These results are fully consistent with the available results on the trp repressor-DNA interaction and confirm that all predicted hot spot amino acid residues are clustered in a spatial shape which fits the helical structure of the bound DNA segment.

Gene expression is regulated not only during development but also during the lifetime of differentiated cells. Abnormalities in gene expression can result in a variety of different disease states and include oncogenic transformation of cells. Genes are regulated by the binding of regulatory proteins to certain responsive elements of noncoding DNA (e.g. operators, promoters or enhancers). This protein-

DNA binding is highly specific and can produce both positive and negative regulatory effects on gene expression. The process of gene regulation has been studied extensively, but the submolecular parameters responsible for the selective recognition between regulating proteins and DNA are still not well defined. The RRM has now been used as a possible approach to defining these parameters. In earlier investigations [24] it was demonstrated that different groups of DNA regulatory sequences are characterised by specific RRM frequencies, whilst proteins which bind to and regulate these DNA sequences (repressors, RNA polymerases, glucocorticoid receptors and homeoproteins) are characterised by the same RRM frequency as the related DNA sequences. These results suggest that the characteristic RRM frequency can be considered as one of the parameters which define specific protein-DNA recognition. Here, we expand on these concepts for the definition of the amino acid residues and the 3D shape requirements of the protein binding site which are critical for this recognition. For these studies, the trp repressor was used to validate the approach because the trp repressor crystal structure and DNA binding sites are well known [84].

The trp repressor characteristic frequency was identified at $f = 0.099 \pm 0.08$ (S/N = 198) from the cross-spectral analysis (Fig. 6.6) of 25 different repressors. Once this RRM characteristic frequency had been defined, it was possible by an inverse RRM procedure to identify the amino acid residues in the trp repressor protein sequence which predominantly contribute to this specific characteristic frequency. These hot spot amino acid residues were identified at the sequence positions: Gly 52, Thr 81 and Thr 83. They were then positioned in the 3D structure of the trp repressor dimer-operator complex (Fig. 6.7) and are highlighted by CPK structures in yellow. It is evident from Figure 6.7 that all hot spot amino acid residues predicted for the monomeric form of the trp repressor are clustered in the topographical space where actual interaction occurs. Two predicted hot spot residues, Thr 81 and Thr 83, have previously been identified from mutagenesis studies [84] to form critical operator recognition contacts. The third predicted hot spot residue, Gly 52, is in close proximity to these amino acid residues which surround the L-tryptophan binding site and is involved in the mechanism by which the binding of L-tryptophan triggers repressor activity [84]. Furthermore, it can be observed from Figure 6.7 that the predicted hot spot amino acid residues form a shape which is in complete complementary accordance with the shape of the interacting DNA. These results with the trp repressor, which is a well-characterised example of a DNA binding pro-

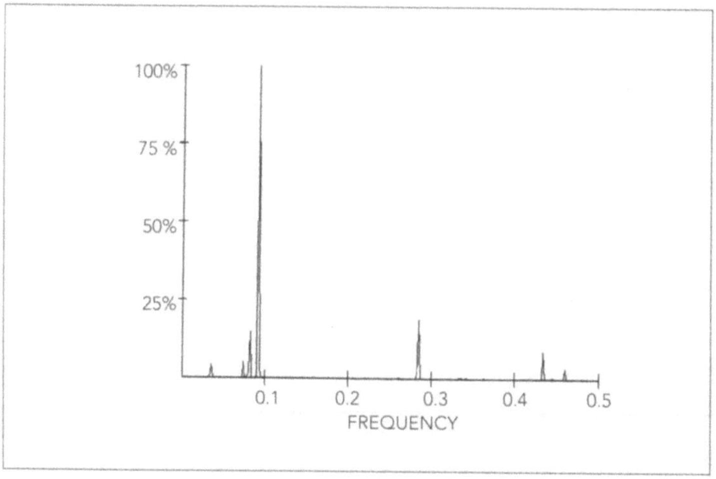

Figure 6.6 Consensus RRM spectrum of 25 repressors
Common characteristic frequency is at $f_1 = 0.099 \pm 0.008$.

tein, validate the ability of the RRM procedure to predict the location of amino acid residues within the primary structure of a repressor protein which are critical for protein-DNA recognition.

These investigations extend the utility of the RRM procedures to DNA binding proteins. It was shown with the trp repressor that the predicted hot spot amino acid residues are among residues which form critical recognition contacts with the operator, or which are related to the L-tryptophan binding site. Furthermore, these predicted hot spot amino acid residues are found to have a spatial distribution in the 3D structure of the trp repressor which forms an interactive surface and shape complementary to the double helix structure of DNA.

Protein-DNA interactions: frequency requirements

Investigations of characteristic resonant recognition frequencies and their biological significance has demonstrated that proteins and their protein or DNA targets exhibit the same characteristic frequency. Thus, it can be proposed that this characteristic frequency is associated with selective recognition between interacting biomolecules. It has been postulated that these characteristic frequencies are electromagnetic in nature. Thus, the requirements for specific protein-DNA com-

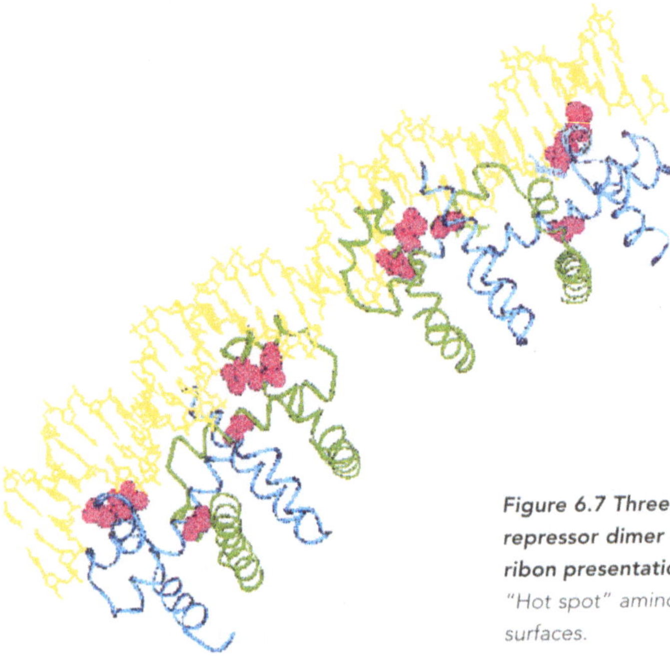

Figure 6.7 Three dimensional structure of trp-repressor dimer – operator complex shown in ribon presentation

"Hot spot" amino acids are highlighted by CPK surfaces.

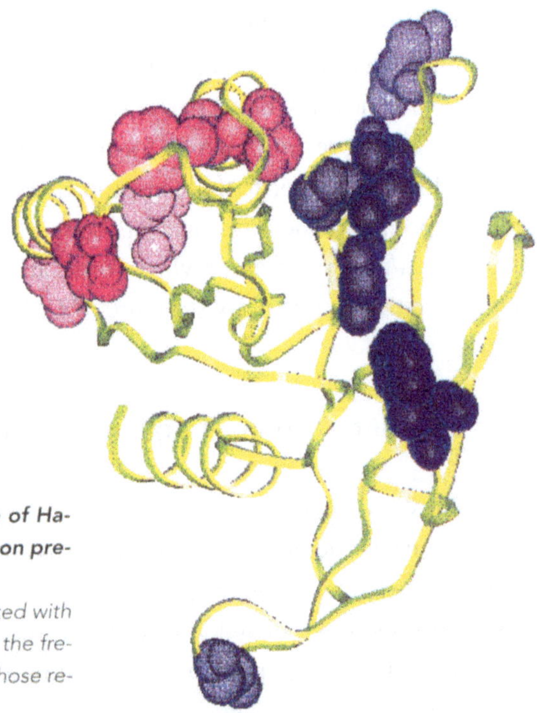

Figure 6.9 Three-dimensional structure of Ha-ras oncogene product p21 shown in ribon presentation

The "hot spot" amino acids are highlighted with CPK surfaces in pink for those related to the frequency f=0.0322 ± 0.004 and in blue for those related to the f=0.0537± 0.004.

plex formation can be considered to involve the coupling of the characteristic electromagnetic oscillations of the protein and its target (DNA).

DNA is a long molecule having different functional segments. These segments are activated by selective binding of DNA binding proteins to the regulatory DNA segments. It is mostly accepted that requirements for this binding are structural. The 3D structure of DNA is usually helical or extended and is not dependent on the moledule's function; thus, structural requirements are more or less the same for any DNA binding protein. However, this binding is very selective, and different proteins can bind different segments. Hence, not only is the 3D structure of protein predominant in determining binding selectivity, but one or more additional parameters may be needed as well. Here we present the hypothesis that RRM frequency may be the critical parameter for selectively binding proteins to DNA.

Using the example of operators, repressors, promoters and RNA polymerase, we showed earlier that DNA segments and corresponding binding proteins have the same characteristic RRM frequency. Consequently, it may be that protein-DNA recognition and binding follow the same principle as protein-protein recognition and binding; i.e., binding participants have the same characteristic RRM frequency.

Frequency requirements for protein-DNA binding are explained here using the example of enhancers and enhancer binding proteins. Enhancers are DNA sequences found in eukaryotic cells and virus-infected eukaryotic cells which have a role in gene expression. They enhance transcriptional activity in both orientations (5'–3' and 3'–5') and can be placed upstream or downstream, even several kilobases from a gene whose transcription they affect. They affect heterologous promoters and show specie specific activity [67, 68].

Two groups of proteins are proposed as enhancer binding and activating: homeobox proteins and glucocorticoid receptors. Homeoproteins are involved in body development, in particular in *Drosophila* segmentation. Thus, they are proposed to be DNA – possibly enhancer – binding proteins. The glucocorticoid receptor is known to stimulate the expression of a number of genes, including MTV, mouse and human metallothionein, growth hormone and α2u globulin genes [85]. DNA sequences mediating transcriptional stimulation of several of these genes have been located and shown to be discrete genomic regions upstream of the transcriptional start site. Some of these sequences appear to exert their actions independently of orientation and position. The steroid receptor complex has been shown to bind to some of these regulatory sequences [85]. The glucocorticoid-responsive sequence has the ability to maintain promoter stimulation despite alteration in its position and orientation [18].

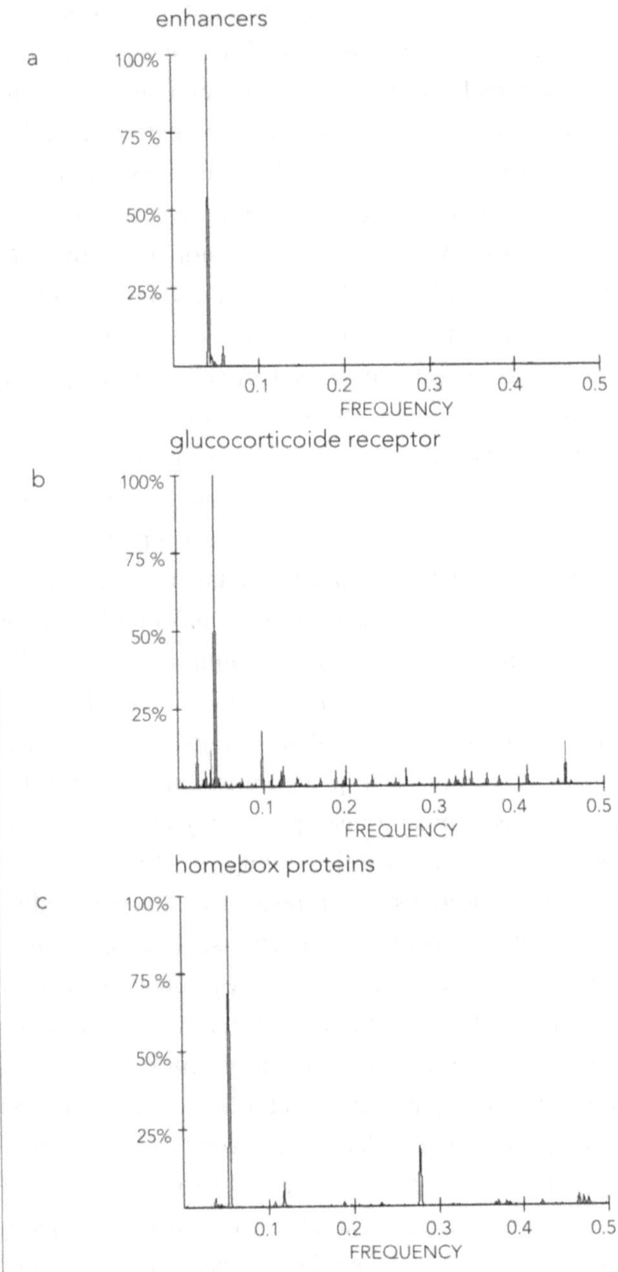
enhancers
a
100%
75 %
50%
25%
0.1
0.2
0.3
0.4
0.5
FREQUENCY
glucocorticoide receptor
b
100%
75 %
50%
25%
0.1
0.2
0.3
0.4
0.5
FREQUENCY
homebox proteins
c
100%
75 %
50%
25%
0.1
0.2
0.3
0.4
0.5
FREQUENCY

This suggests a mechanism of action resembling that of enhancer sequences.

As described above, cross-spectral analysis was applied to 10 enhancers from different origins. The common characteristic frequency for all analysed enhancers was found to be f = 0.0488 ± 0.024 with S/N = 467 [27, 28] (Fig. 6.8a).

Since previous observations have shown that activation of an enhancer requires specific binding proteins, the RRM was used to analyse possible enhancer binding proteins. The cross-spectral function for four glucocorticoid receptors (human α, β, rat and mouse) resulted in a spectrum with the highest peak representing the most common frequency component. This frequency component was found at f = 0.0439 ± 0.001, and it overlaps with the enhancer frequency as shown in Figure 6.8b. The same analysis with homeoprotein domains revealed that these proteins have one common characteristic frequency at 0.0459 ± 0.001 (S/N = 100) (Fig. 6.8c). The characteristic frequency of homeopathic domains is within the range of enhancer frequency, implying that their function may correlate with the role that enhancers play in gene regulation. However, the characteristic frequency of homeoproteins does not overlap the glucocorticoid receptor characteristic frequency, which seems to indicate that these two groups of proteins have different functions, though both can bind and activate the same regulatory elements on DNA.

New possibilities: *ras*-DNA interaction

Investigations of cellular and molecular mechanisms of uncontrolled cell proliferation associated with transformation have resulted in the identification of a large number of genes (oncogenes), the products of which are involved in cell neoplasia. The oncogenic family of proteins is generally involved in regulation of cell proliferation through regulation of DNA transcription. This regulation can be achieved indirectly via different signal pathways as is the case with tyrosine kinase oncogene products, e.g. *src, ros, fps, fes* and *yes*, or directly by protein-DNA binding, as is the case with nuclear proteins like *myc, fos* and *myb*. There is also a group of oncogene products including *ets*-, *sfi*- and *ras*-related proteins, for which *in vivo*

◄ ***Figure 6.8***
(a) Consensus RRM spectrum of enhancers. Common characteristic frequency is at f = 0.0488 ± 0.024. (b) Consensus RRM spectrum of four glucocorticoida receptors (human α, β, rat and mouse). Common characteristic frequency is at f = 0.0439 ± 0.001. (c) Consensus RRM spectrum of homeoprotein domains. Common characteristic frequency is at f = 0.0459 ± 0.001.

biological functions and pathways of action still have to be fully defined. The Ha-*ras* oncogene family (the p21 proteins) are known to function *in vitro* as GTP binding proteins involved in signal transduction pathways, as well as in control of DNA synthesis, cell transformation, proliferation and differentiation [75]. Although it is known that the cellular, proto-oncogenic form of p21 is involved in the control of cell proliferation, while the oncogenic form is additionally involved in cell transformation and differentiation, the exact mode of action and signal pathways of the Ha-*ras* p21 proteins (both cellular, proto-oncogenic and oncogenic) remain unknown.

As described in Chapter 5, the p21 *ras* oncogene products share a similar ability to transform cells, as do a number of other oncogenic proteins. Thus, a whole set of 46 different oncogenes was analysed using the RRM, and the common characteristic frequency component was identified for the set at the frequency f = 0.0322 ± 0.004 with S/N = 468 (Fig. 5.12). According to the RRM, this result suggests that this common frequency characterises a common biological behaviour of this group of transforming proteins. From the cross-spectral function (Fig. 3), a second, less prominent peak is also observed at frequency f = 0.0537 ± 0.004 with S/N of 83, which is also associated with a structure-function property common to all these cellular proto-oncogene and oncogene products.

With reference to previous investigations [27, 28], it can be observed that the frequency f = 0.0322 ± 0.004 is very close (that is, within the calculation error) to the frequency characteristic of proteins capable of DNA binding and regulation via enhancers. Thus, this frequency may characterise properties of p21 which are similar to DNA binding and regulation of DNA replication, as manifested by other nuclear proto-oncogenes like *fos, jun, myc* and *myb*. Based on the cross- correlation analysis of the Fourier spectra of the Ha-*ras* family of oncogenes with the spectra of the *fos, jun, myc* and *myb* oncogenes, the results indicate that these gene products do share a common frequency characteristic which is associated with their functional potential for DNA binding. Using RRM procedures, the amino acid residues which predominantly contribute to these two specific characteristic frequencies, hot spots, were defined for the Ha-*ras* p21 oncogene products. The hot spot amino acids related to the frequency f = 0.0313 were found Gly 10, Gly 12, Gly 13, Gly 15, Phe 28, Gly 48 and Ser 65. As discussed earlier, the analysis indicated that the hot spot amino acids at positions Gly 10, Gly 12, Gly 13, Gly 15 and Phe 28 contribute more to the characteristic frequency for p21 oncogenic and GTP-binding function than the amino acids at the positions Gly 48 and Ser 65. Significantly, amino acid residues

Gly 12, Gly 13 and Phe 28 have been found to be among the GTP active site residues [32], while Gly 10 and Gly 15 are spatially clustered around the active site as is evident from the X-ray crystallographic structure of p21 (Fig. 6.9, see page 104).

The hot spot amino acid residues related to the DNA regulating frequency (f = 0.0537) were found at Ser 89, Phe 90, Ser 106, Ser 127 and Gly 138. In Figure 6.9, the locations of both groups of these amino acids are presented in the 3D structure of the Ha-*ras* p21 oncogene product as determined by X-ray crystallography [28]. Hot spot amino acids related to the frequency common for oncogenic GTP-hydrolysis function (the "oncogene frequency") are highlighted with CPK surfaces in pink, while the hot spots related to the DNA regulating frequency (the "enhancer frequency") are highlighted with CPK surfaces in blue. According to the X-ray crystallographic structure of the guanine-nucleotide binding domain of p21 [76], the guanine base can be positioned in the tertiary structure of the p21 protein, as shown in Figure 6.9. From the data presented, the following conclusions can be reached:

1. Hot spot amino acids related to two different functional responses are positioned in two distinct topological clusters in the 3D structure of the p21 protein.
2. Hot spot amino acids related to oncogenic function are not sequentially linked in the protein sequence but, nevertheless, form a cluster in the tertiary structure which is positioned in and around the guanine nucleotide binding site. This topological region specifically involves the relevant hot spot amino acid residues: Gly 10, Gly 12, Gly 13, Gly 15 and Phe 28.

To further evaluate these predictions of the structure-function properties of the Ha-*ras* p21 oncogene product, we have examined the spatial disposition of the predicted hot spot amino acids in the 3D structure of the p21 protein in relation to the complementary binding with DNA. Although the prediction of these hot spots has been derived from the analysis of the ability of GTP to bind to the p21 guanine binding site *in vitro*, our results are clearly applicable to guanine, which is incorporated in DNA as well. As can be seen from the structural projection shown in Figure 6.9, all hot spots related to the oncogenic function (blue CPK's) form a surface very similar to the imprint of the helical shape of DNA. To provide further support for this concept, the spatial features of this surface have been compared in terms of the complementary fit with the shape of the 3D structure of a helical fragment of DNA. When the DNA helix is superimposed on the structure of p21 with the guanine nucleotide positioned in the GTP-binding site, the spatial shape of the predicted hot spot topographic region of p21 exhibits a very high complementary fit (on the average a distance of 4.3 Å with SD of 1.5 Å) to the shape of the superimposed DNA fragment [65]. These results with p21, analogous to the results obtained with the trp repressor, indicate that the p21 protein has topological features which would permit binding to particular DNA segments. Furthermore, the spatial arrangement of the predicted set of hot spot amino acids related to the oncogenic function accurately define the geometry of this binding site.

A commonly accepted model for the function of the Ha-*ras* p21 oncogene product involves its action as a plasma membrane-associated protein by transmitting the signal(s) necessary for growth promotion through pathways similar to these of other G-proteins [86]. According to the results presented, the hot spot amino acids crucial for the oncogenic behaviour of the p21 protein can be identified by the use of the RRM informational algorithm and molecular graphic procedures. These amino acids are associated with the guanine binding site and are known from mutagenesis research [77] to be associated with cell transformation. Furthermore, the results indicate that these hot spots form a spatial shape in the tertiary structure of the p21 protein which, as a complementary surface, corresponds to the helical shape of DNA segments. In the light of these observations and the independent experimental findings that the oncogenic form of the Ha-*ras* p21 protein can directly activate the VL30 transcriptional element [87], the present study suggests that p21-related proteins can act as DNA binding proteins.

DNA binding is not necessarily synonymous with regulation of DNA transcription. However, our results indicate that the second characteristic frequency at $f = 0.0537$ (indicative of the so-called enhancer function) of p21 is associated with a biological process involved in the regulation of DNA replication. This result is in accordance with investigations which previously examined other DNA regulating proteins, including glucocorticoid receptors and homeodomain proteins [27]. From this point of view, the hot spot amino acids identified and related to oncogenic function denote an active site for DNA binding, while the hot spot amino acids related to the enhancer function may denote an active site for binding to an additional activator protein. Previous studies [76] have shown that the site of interaction of the GTPase activating protein (GAP) with p21 involves residues 32 to 40, encompassing part of the so-called l_2 loop and part of the b-strand b_2. From these investigations, it can be concluded that the second cluster of hot spot amino acids associated with the enhancer function denotes an additional putative binding site for another protein regulator of p21 action. Further, all of these residues are within the C-domain structure of p21 and are spatially encompassed within the loop l_7, the kinked helix a_3 and the helix a_4.

These results are also in accord with concepts of gene-activating proteins [88], which propose that DNA activation is not solely controlled by DNA binding proteins but in some cases involves an additional activator protein bound to the DNA binding protein.

Thus, the results with the well-characterised example of the trp repressor, presented above, validate the composite approach involving RRM techniques and molecular graphics for evaluating protein-DNA interactions. These results with the less well characterised example of the Ha-*ras* p21 transforming protein have significant predictive value. They provide the first and somewhat surprising indication that p21 proteins, besides their ability to bind free guanine *in vitro*, have the structural ability to bind guanine incorporated into the DNA sequence as well. As p21 also shares the frequency characteristic of the so called enhancer function, which previously was found to be a characteristic of DNA regulation proteins via enhancers, our results indicate that p21-related proteins may thus function *in vivo* also by regulating DNA transcription. This behaviour would be in accord with evidence confirming that p21 can indeed activate the VL30 transcriptional element [88], induce neuronal differentiation [89] and act as part of a signal transduction pathway in cells. Moreover, the translocation of GTP-binding proteins from the cell periphery to the nucleus has been detected with Balb/c 3T3 fibroblasts under stimulation by phorbol dibutyrate, insulin or EGF [90]. Thus, the results with Ha-*ras* p21 transforming proteins can be discussed in the light of a possible p21 translocation to the nucleus when it is complexed with one of two recently identified GAPs [86]. This process would be analogous to several known examples of plasma membrane or cytolic proteins which are translocated to the nucleus and which act as regulators of gene transcription when they are activated by steroid hormones, nucleotide binding or kinases [91]. DNA binding and regulation could thus represent an additional integrated mode of action associated with the oncogenic properties of p21-related proteins.

7 Bioactive Peptide Design

Once the characteristic frequency of the particular biological function/interaction is identified, it is possible to design an amino acid (or nucleotide) sequence that has only this spectral characteristic. We propose that this sequence designed *de novo* will have only the desired biological function related to the chosen characteristic frequency. The procedures and experimental results for the design of this bioactive sequence are described below.

Following the determination of the characteristic RRM frequency and corresponding phases for particular biological functions, it is possible to design amino acid sequences which will have those spectral characteristics only. It is expected that these peptides will exhibit the desired biological activity. The strategy for the design of such defined peptides is as follows:

1. Determination of the characteristic unique frequencies for the specific biological function or interaction analysed by multiple cross-spectral analysis for the group of protein sequences with this corresponding biological function as described above.
2. Definition of the characteristic phases at the characteristic frequencies for the particular protein which is chosen as the parent for agonist/antagonist peptide design.
3. Derivation of a numerical sequence from the characteristic frequencies and phases. This is done by summing sinusoids of the particular frequencies, amplitudes and phases. The length of the numerical sequence is defined first by appropriate frequency resolution, which allows different characteristic frequencies to be separated, and, second, by the peptide length desired.
4. Determination of the amino acid that corresponds to each element of this new numerical sequence can then be achieved from tabulated values of EIIP or other appropriate amino acid coefficient parameters. Peptides obtained in

 ©1997, Birkhäuser Verlag Basel

this way will have the desired spectral characteristics and are purported to have the corresponding biological activity. The application of the above strategy is now presented.

7.1 Procedures: example of design of glucagon analogues

Detailed procedure used for *de novo* design of bioactive peptides are presented here with glucagon agonists as an example.

The aim was to design shorter peptides of glucagon having the same biological function. Application of RRM analysis to 13 different glucagons from 18 species and 7 different glucagon-like peptides revealed a single prominent peak in the cross-spectral function at the frequency $f = 0.3242 \pm 0.034$ ($S/N = 70.54$). The frequency component at $f = 0.3242$ was common to all glucagon-related polypeptides analysed, and thus could be considered the consensus frequency characteristic of these polypeptides for a common biological property, for example, the potential to stimulate glycogenolysis.

The next step was to define the active site of the human glucagon. Having identified the characteristic frequency of glucagons, it was possible to identify amino acids in the human glucagon sequence that are major contributors to the characteristic frequency. The locations of these predicted amino acids are within the known functionally important regions of the human glucagon sequence: three of the predicted amino acid residues (Gly 4, Thr 5 and Thr 7) are located within the N-terminal region, known to be critical for transduction of the biological message, whilst the other two amino acids residues (Phe 22 and Thr 29) are located at the C-terminal of human glucagon, previously shown [92-94] to be involved primarily in receptor recognition. This correlation between RRM predictions and experimentally derived data provides additional evidence that the concepts implicit in the RRM model can be employed as an aid to the prediction of the contribution which a specific amino acid residue may make to the global arrangement of a biorecognition site in a polypeptide.

In order to examine the spatial disposition and topology of these predicted amino acid residues, an appropriate solution structure for human glucagon was needed. A currently accepted model for human glucagon in solution is the model

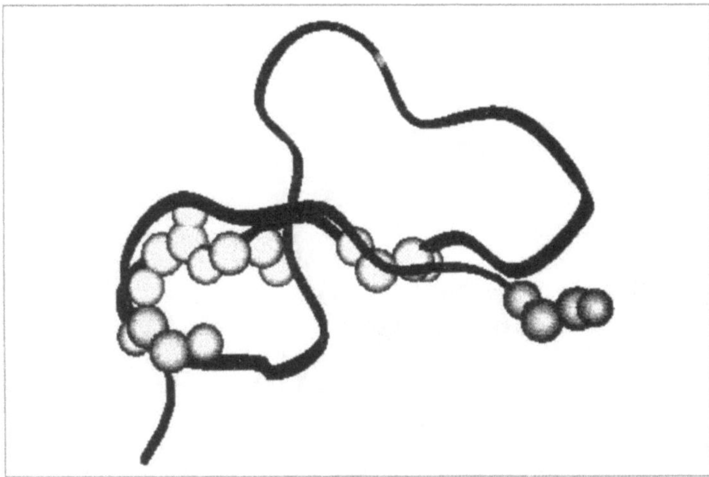

Figure 7.1 Predicted 3D structure of the human glucagon molecule in ribbon presentation

Predicted hot spot amino acids are highlighted with CPK surfaces and are clustered in a space around the cleft of the active site.

of Korn and Ottensmeyer [95], which is essentially based on solution measurements (i.e. CD (Circular Dichroism), fluorescence quenching and NMR) as well as Chou-Fasman [7] secondary structure predictions. Our investigations suggested a modified model of the structure of human glucagon in solution. The model resulted from conformational simulations using molecular mechanics and dynamics [10-12]. This stable conformation represents a local energy minimum which can be achieved under conditions at 300 K, i.e. energy conditions which may approximate those within the living cell. This simulated human glucagon conformation agreed well with previous experimental observations as well as with the Korn-Ottensmeyer model in terms of the compact overall structure with the N- and C-termini crossing each other and the general shape of the backbone structure [95]. Consequently, this simulated structure was used as a model of the tertiary structure of human glucagon in solution in order to investigate the spatial disposition of the functional key amino acid residues predicted by the RRM. In the case of human glucagon, these amino acids were clustered along one face of the structure of the globular polypeptide, forming an arch across the N- and C-terminal regions and the hydrophobic cleft (Fig. 7.1). This result strongly supports earlier conclusions reached with a number of other polypeptides and proteins, that such key amino acid residues predicted by the RRM are located in spatially clustered regions in the tertiary struc-

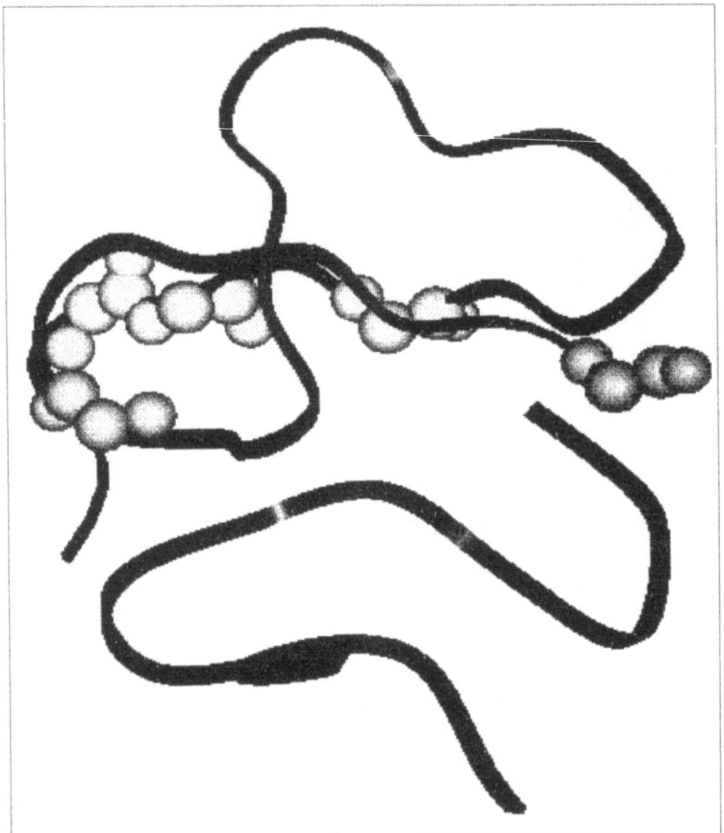

Figure 7.2 Ribbon structures of glucagon (as predicted by molecular-modelling procedures) and 18-mer peptides designed with the RRM to have the glucagon characteristic frequency

The designed peptide has a structure very similar to the glucagon active site, which is highlighted by CPKs. There is no significant homology between designed peptide and glucagon.

ture of the polypeptide or protein and denote a specific biorecognition or interaction site.

Having determined a glucagon characteristic frequency, it was possible to design a variety of polypeptide analogues having only this spectral characteristic [24, 25, 35]. In particular, several polypeptides were designed to satisfy the criterion that their RRM spectra should exhibit only the frequency $f = 0.3242$. It is important to note that no significant homology exists between these derived peptides and human glucagon, whilst all these peptides show very high mutual sequence homology. The structures of these peptides were then subjected to molecular mechanics and dynamics simulations using the same procedures as for human glucagon. Simulated structures of these peptides were compared in terms of superimposition of their backbones onto the structure of the human glucagon interaction site, encompassing the cluster of key amino acid residues (Gly 4, Thr 5, Thr 7, Phe 22 and Thr 29). Goodness of fit was determined from the root mean square deviation (RMS) of the peptide backbone from that of the corresponding interactive region of human glucagon.

The best fit, RMS = 1.98 Å, for peptides in the range of 15-*mer* to 20-*mer* was found for the 18-*mer* peptide with the following primary sequence, LCRIQDGQDEWDPYDHKD. The predicted 3D structure of the designed 18-*mer* peptide is comparable in backbone topography to the structure of the glucagon active site, while it is not similar in primary structure to any sequentially linked fragment of human glucagon (Fig. 7.2). Since the biorecognition sites of a protein are usually composed of amino acids which are not contiguously linked, the folded structures of predicted peptidic analogues should mimic the overall shape of the binding site rather than have primary structures which are simply homologous to the sequentially linked fragments of the protein under examination.

Finally, additional studies are necessary to characterise the biological properties of synthetic peptide analogues developed from the composite computational approach described above. Such biological evaluations have thus far been successfully applied in a few examples, including the design of immunological analogues of the HIV envelope protein [96] as well as FGF antagonist [34].

7.2 NGF analogue design and possible applications

Procedures for a bioactive peptide design, described in the previous chapter, have been applied to the design of peptides analogous to nerve growth factors. A wide variety of applications are discussed.

NGF is a factor for maintaining and differentiating sensory and sympathetic neurones [97]. It is isolated first from snake venom and then from mouse submaxillary glands as a 7S complex containing three subunits: α, β and γ. Activity resides in the β chain, which is a dimer of two identical, 118-*mer* chains. Receptors for NGF are present on a variety of normal sympathetic and sensory neurones as well as in normal and neoplastic chromatin cells.

The RRM was applied to a group of eight NGFs, including snake venom and mammalian factors, and the analysis revealed à prominent common frequency characteristic at $f = 0.404 \pm 0.008$ with S/N = 192 (Fig. 6.3a) [26]. In order to find out whether this frequency is characteristic of NGF-receptor recognition, NGFs were compared with their receptors, and an even more prominent peak (S/N = 282) occurred at the same frequency (Fig. 6.3b). Phases at the characteristic frequency were found to be $\varphi = -0.34$ rad and $\varphi = 2.62$ rad for human NGF and its receptor, respectively. The phase difference was $\Delta\phi = 3.96$ rad, close enough to π, which is in accordance with the RRM postulate that proteins and their receptors have opposite phases at the characteristic frequency.

Once we had identified the human NGF characteristic frequency and phase, we were able to design the 18-*mer* peptide, DKYRLDHWRNDHQFGDPQ [26]. As was shown in experiments with FGFs [34] and HIV env [24, 96], described later, it is expected that such a designed peptide will have the same function and receptor recognition as native protein. Design of synthetic, shorter NGF means easier and less costly production of factors that can induce nerve growth within an organism or cell culture.

7.3 Design of FGF analogues and application in limiting solid tumor growth

The proliferative and angiogenic potential of FGF-b has been documented during embryogenesis, vascularisation of the ovary and related reproductive tissues, in pathological states such as wound healing, tissue repair and tumour development and expansion [97]. The angiogenesis of solid tumors is a of function of FGF which is of great interest. The design and synthesis of highly specific peptide analogues capable of interacting with FGF receptors may provide a therpeutic mode for the prevention and treatment of solid tumour angiogenesis. In these studies RRM was used to define frequency characteristics of FGF's responsible for mitagenic activity and receptor recognition.

The multiple cross-spectral function of the three acidic and four basic FGFs (Fig. 7.3a) reveals one prominent peak at the frequency f_1 = 0.4521 ± 0.005 (S/N = 120). In order to determine whether this frequency was relevant for FGF receptor recognition, a further comparison with two FGF receptors, flg and bek was carried out. The resulting multiple cross-spectral function shows that frequency f_1 is even more prominent (S/N = 215), while the other frequency components were diminished (Fig. 7.3b). From these data, it can be concluded that frequency f_1 is relevant for recognition between FGF's and their receptors. The other frequency component, f_2 = 0.255 ± 0.002, was found to characterise the biological activity of competence growth factors, including FGFs (Fig. 7.3c). Thus, in the case of the FGFs it was possible to distinguish between receptor recognition and growth-promoting activity in terms of different characteristic frequencies. Consequently, peptides designed to have receptor binding characteristics without growth-promoting activity incorporate a frequency component at f_1 = 0.4521 and omit the frequency component at f_2 = 0.256. Using the peptide design module of the RRM, a number of peptides satisfying the frequency conditions were designed with lengths of 10 to 24 amino acids. The procedure was the same as that described for glucagons.

Molecular-modelling techniques were applied to the new peptides, with the aim of identifying their stable structures. The proposed structures were then compared with the 3D structure of the FGF 97-120 fragment, which is known to be a receptor binding domain. Although no sequence homology exists between the designed peptides and the FGF 97-120 fragment, a high structural similarity was detected. On the basis of RMS deviations between the predicted backbone structures, a peptide 16 amino acids in length (FGFRRM16) was found to be the one most similar (RMS = 1.37Å) to the FGF 97-120 fragment. The predicted 3D structures of FGF fragments 97-120 and FGFRRM16 are shown in Figure 7.4. The proposed FGFRRM16 amino acid sequence is MWYRPDLDERKQQKRE

The designed peptide FGFRRM16 was synthesised, and its function wastested experimentally. Fibroblast cultures were used for testing the bioactivity of FGFß peptide antagonists, since FGF-β exerts stimulatory effects on DNA synthesis and the proliferation of these cells *in vitro* [21]. These observations were confirmed in both [^{3}H]thymidine incorporation and cell proliferation bioassays with a dose-dependent stimulation of tracer uptake and proliferation of fibroblasts in culture [34, 25]. The peptide itself had no effect on [^{3}H]thymidine uptake or the proliferation of fibroblasts in culture. However in the presence of a stimulatory dose of FGF-β (25 ng/ml), peptide FGFRRM16 antagonised the actions of FGF-β, inhibiting trac-

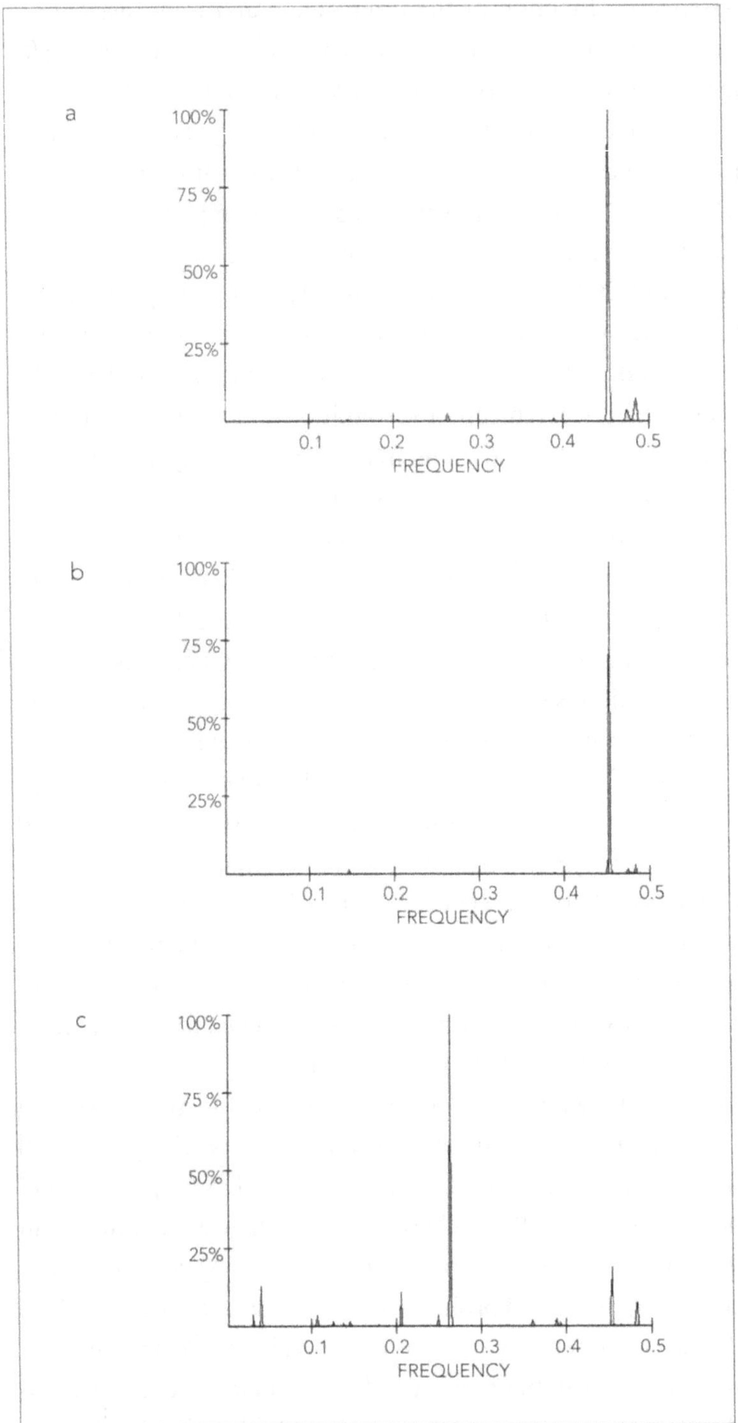
a
100%
75 %
50%
25%
0.1
0.2
0.3
0.4
0.5
FREQUENCY
b
100%
75 %
50%
25%
0.1
0.2
0.3
0.4
0.5
FREQUENCY
c
100%
75 %
50%
25%
0.1
0.2
0.3
0.4
0.5
FREQUENCY

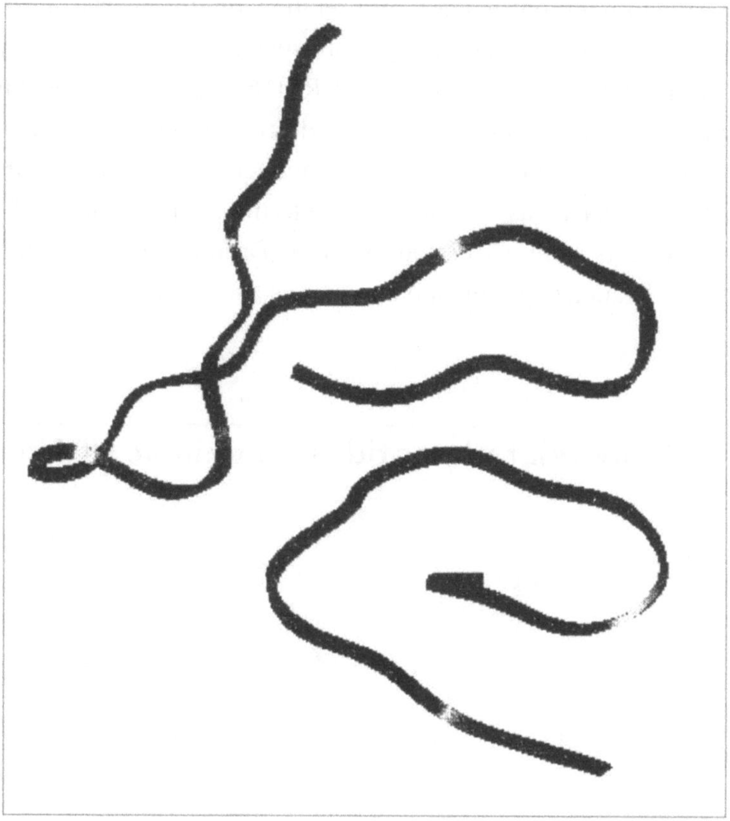

Figure 7.4 Ribbon structures of the (up) FGF active site (97-120 fragment) and (down) FGFRRM16 peptide
Although there is no sequence homology between them, their 3D backbone shapes are comparable.

◀ **Figure 7.3 Characteristic RRM frequencies of FGFs**
(a) Cross-spectral function of five FGFs (three acidic: human, bovine, rat; and two basic: human, bovine). The prominent peak is at 0.4512 ± 0.005 with S/N = 121. (b) Cross-spectral function of five FGFs with two FGF receptors (bek and flg). The same frequency peak is prominent but has a higher S/N ratio, 215. (c) Cross-spectral function of five FGFs, seven interleukins-1, and two TGFs. These proteins all belong to the wider functional group of competence growth factors. The common frequency component was found at the frequency 0.263 ± 0.002, with S/N = 206.

er incorporation and the proliferation of fibroblasts *in vitro*. Cell proliferation bioassays were consistently more sensitive than were DNA synthesis bioassays, with doses of 0.5 mg/ml of FGFRRM16 antagonising and 2 mg/ml of FGFRRM16 totally blocking FGF-β-stimulated proliferation of fibroblasts. It is important to note that while DNA synthesis continued at 2 mg/ml of FGFRRM16, this replication of DNA was not translated in terms of proliferation of fibroblasts. These experiments were performed by Dr. A. Drummond (Monash University, Australia). and detailed experimental procedures and results are presented in reference 34.

7.4 HIV-related peptides that mimic HIV immunogenicity

The interaction between HIV virus envelope proteins and CD4 cell surface antigen has a central role in the process of virus entry into the host cell. Thus, blocking the interaction between the envelope glycoproteins and CD4 surface antigen, known to be the HIV receptor, should inhibit infection [98]. The RRM model was used to analyse these interactions and consequently to design peptides capable of mimicking the immunorecognition function of the HIV $gp160^{env}$ protein.

Prior to designing peptide analogues capable of functioning as HIV-CD4 interaction inhibitors, we needed to define the characteristics of the $gp160^{env}$ proteins crucial for their interaction with CD4. Despite the high sequence variability between different isolates, it can be said that at least one specific characteristic has to be common for all the different isolates. Envelope proteins from 21 HIV isolates (lav1/bru, hxb2, nl43, sf2, sc, mn, rf, wmj2, cdc451, ny5, jh3, brva, eli, mal, z6, z2z6, z3, z321, jy1, ndk, oyi) were analysed using the RRM with the aim of defining a common frequency characteristic [25, 96]. Only one prominent peak was obtained in the cross-spectral function at the frequency $f_1 = 0.1855 \pm 0.001$ with S/N = 484 (Fig. 7.5a) This frequency component was common to gp160 glycoproteins for all HIV isolates analysed. To validate the idea (in accordance with the RRM) that this frequency characterises the common biological behaviour of all proteins analysed, i.e. recognition and binding to CD4, the CD4 molecules (human and mouse) were compared with the cross-spectral function of 25 previously obtained $gp160^{env}$ proteins. The prominent peak at the same frequency, $f_1 = 0.1855$, occurs but the S/N is significantly higher, implying that CD4 molecules also share the same frequency characteristic [25, 96]. This is completely in accordance with previous findings [27, 29, 33] in a number of examples, namely that protein ligands and their receptors (irrespective of whether they are protein or DNA targets) share the same characteristic frequency, which defines their mutual recognition.

Regions from gp120 which are crucial for binding to CD4 [99], 44 amino acids in length, from 11 HIV isolates (lav, arv2, eli, ny5, wmj2, mal, z3, 3cg, cdc451, hxb2, hat) were also analysed. Cross-spectral analysis revealed that those regions share the same characteristic frequency with the gp160^{env} protein, but they also have their own prominent characteristic at the frequency $f_2 = 0.2188 \pm 0.022$ (Fig. 7.5b).

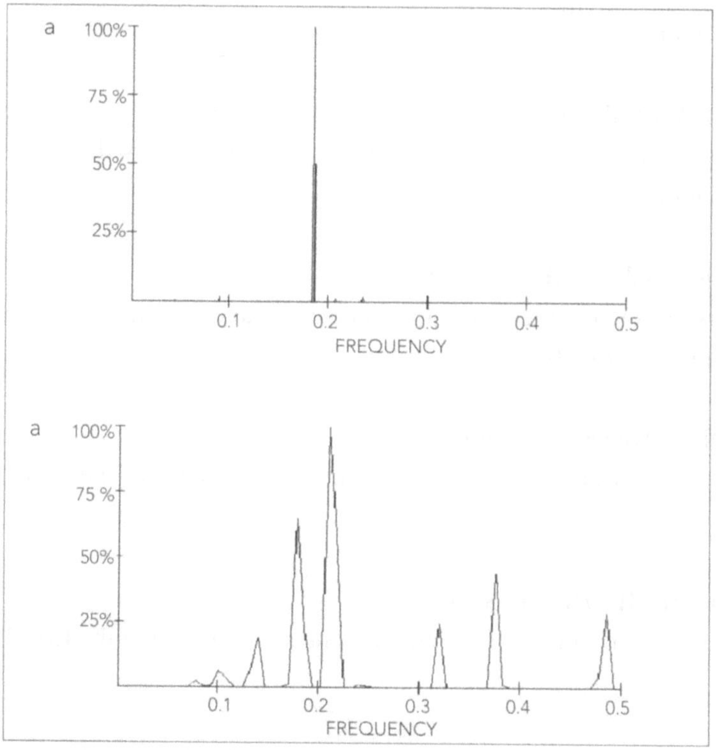

Figure 7.5 HIV envelope characteristic frequencies

(a) Multiple cross-spectral function of gp160^{env} from 21 HIV isolates (lav1/bru, hxb2, nl43, sf2, sc, mn, rf, wmj2, cdc451, ny5, jh3, brva, eli, mal, z6, z2z6, z3, z321, jy1, ndk, oyi). Prominent peak obtained at the frequency 0.1855 ± 0.001 with S/N = 484. (b) Multiple cross-spectral function of regions from gp120 which are crucial for binding to CD4, 44 amino acids in length, from 11 isolates (lav, arv2, eli, ny5, wmj2, mal, z3, 3cg, cdc451, hxb2, hat). Cross-spectral analysis reveals that those regions share the same characteristic frequency as gp160^{env} proteins but also have their own prominent characteristic at the frequency 0.2188 ± 0.022.

To provide confirmatory evidence as to whether one or both of these two characteristic frequencies characterise the interaction between the HIV envelope and CD4 antigen, six peptides, each 20 amino acids in length with the following spectral characteristics, were designed [25, 99]:

Peptide A: KQQYYWYAWCQPPQDQLIMD
The spectrum consists of two frequencies, $f_1 = 0.1855$ and $f_2 = 0.2188$, with phases opposite to the phases characteristic of the LAV1 gp120 region.

Peptide B: LKRDQEPMDFHIWDDYLKRD
The spectrum consists of one frequency only, f_1, with its phase opposite to the phase of the LAV1 gp120 region.

Peptide C: TPPTDWLADRHEMDQNKDDK
The spectrum consists of one frequency only, f_2, with its phase opposite to the phase of the LAV1 gp120 region.

Peptide D: DDALYDDKNWDRAPQRCYYQ
The spectrum consists of both characteristic frequencies, f_1 and f_2, with the same phases as these of the LAV1 gp120 region.

Peptide E: DFHIWDDYLKRDQEPMDFHI
The spectrum consists of the frequency f_1, with the same phase as that of the LAV1 gp120 region.

Peptide F: TDWIYDRHEMDQNKDDKNQD
The spectrum consists of the frequency f_2, with the same phase as that of the LAV1 gp120 region.

It is important to note that the frequency f_1 was also found to be significant in the spectra of $p55^{gag}$ and $gp41^{env}$ HIV proteins, while the frequency f_2 was found to be significant in the spectra of $p18^{gag}$ HIV proteins. Thus, it can be anticipated that the designed peptides, besides their reactivity relative to the LAV1 $gp120^{env}$ protein, may have a reactivity relative to the other HIV proteins listed.

To validate the RRM predictions, the peptides designed were chemically synthesised, and their immunoreactivity was tested experimentally. Biological experiments with immunogenicity of synthetic peptides were conducted by Dr. V. Krsmanovic and Dr. J. M. Bicquard (CNRS, France). These investigations were performed initially by evaluating the reactivity and cross-reactivity of all six designed peptides with the corresponding antibodies [96]. These results have shown significant cross-reactivity to the polyclonal antibodies raised against peptides, which share at least one characteristic frequency and phase at this frequency. In contrast, antibodies raised against peptides which do not share a common frequency and phase at this frequency do not show significant cross-reactivity (Tab. 7.1) (for details see ref. 96). These results provide experimental confirmation of the concept that RRM frequency characteristics reflect important parameters associated with biomacromolecular recognition and in particular antibody-antigen recognition.

Table 7.1 Cross-reactivity between antibodies raised against peptides designed with the RRM
The values represent the logarithms of the highest dilutions against which the serum shows significant reactivity. Peptide C could not be analysed in vivo *because this peptide was toxic in experimental rabbits, and peptide F because it was not immunogenic.*

Peptide	anti-A	anti-B	anti-D	anti-E
A	5.118	3.311	2.41	0.9
B	3.612	5.118	2.11	2.7
D	0	0	3.612	2.41
E	0.9	3.612	3.91	5.412

Furthermore, to ascertain whether any of the proposed frequency characteristics are important in determining the interaction between HIV envelope proteins and CD4 surface antigen, the reactivity of antibodies raised against the peptides designed and LAV HIV viral proteins was investigated [25, 96]. These results show clearly that the peptide D is able to induce polyclonal antibodies in rabbits which can be captured by all *gag*-related *env*-related proteins of HIV-1, except the $p15^{gag}$ protein. This cross-reactivity of the anti-D serum is compatible with the existence of the common RRM spectral characteristic(s) which peptide D shares with HIV-1 proteins. If we consider that peptide A has the same frequency components as peptide D but opposite phases at those frequencies, and that anti-A antisera does not react with any of the HIV proteins, we can conclude that phases at characteristic frequencies are also important parameters in biomolecular recognition. Furthermore, peptide E has only one frequency component, f_1, and the phase at f_1 is the same as that of peptide D. Since anti-E antisera does not react with HIV proteins, we deduce that f_1 is not a critical frequency for recognition between CD4 antigens and HIV proteins. Peptide C could not be analysed *in vivo* because this peptide was toxic in experimental rabbits, and peptide F because it was not immunogenic. Nevertheless, at this stage, it is possible to indirectly propose that frequency $f_2 = 0.2188 \pm 0.022$ characterises the recognition between CD4 antigens and HIV proteins. More generally, we can conclude that RRM frequencies together with their corresponding phases are important physicochemical parameters that characterise specific biorecognition processes.

7.5 Activity of HIV-1-inhibiting peptides

Studies on the antiviral activity of peptides from HIV-1 envelope revealed two peptides (DP-107 and peptide 637-666) from gp41 having antiviral activity [100]. The mechanism by which these peptides inhibit HIV-1 infection is still not known. To analyse the inhibitory activity of these peptides, the RRM of protein-protein and protein-DNA interactions was used.

HIV-1-inhibiting peptides which selectively bind to the virus fusion domain [100 should have the same characteristic frequency as HIV envelope proteins with phases opposite at this frequency. As presented above, HIV envelope proteins, gp160 from 19 HIV-1 isolates, were analysed with the RRM. One common frequency was identified at $f = 0.185 \pm 0.001$ (Fig. 7.5a). The same frequency was identified as common to gp41 and gp120, characterising their mutual interaction. When inhibitory peptides were analysed, they both showed a peak at the frequency $f = 0.185 \pm 0.001$. Moreover, for the peptide DP-107, this peak is the most prominent (Fig. 7.6). The average phase for gp120 proteins at the frequency $f = 0.185 \pm 0.001$ was $\varphi_{gp120} = -0.6$ rad, while for gp41, the average phase was $\varphi_{gp41} = -1.5$ rad. The phases at the frequency $f = 0.185 \pm 0.001$ for the inhibitory peptides are found to be $\varphi_{DP\text{-}107} = 1.5$ rad and $\varphi_{637\text{-}666} = 3.0$ rad. Thus phase shifts between peptides and gp120 and gp41 are as follows:

$$\Delta\varphi_{DP\text{-}107\text{-}gp120} = 2.2 \text{ rad}$$
$$\Delta\varphi_{DP\text{-}107\text{-}gp41} = 3.5 \text{ rad}$$
$$\Delta\varphi_{637\text{-}666\text{-}gp120} = -2.7 \text{ rad}$$
$$\Delta\varphi_{637\text{-}666\text{-}gp41} = -1.8 \text{ rad}$$

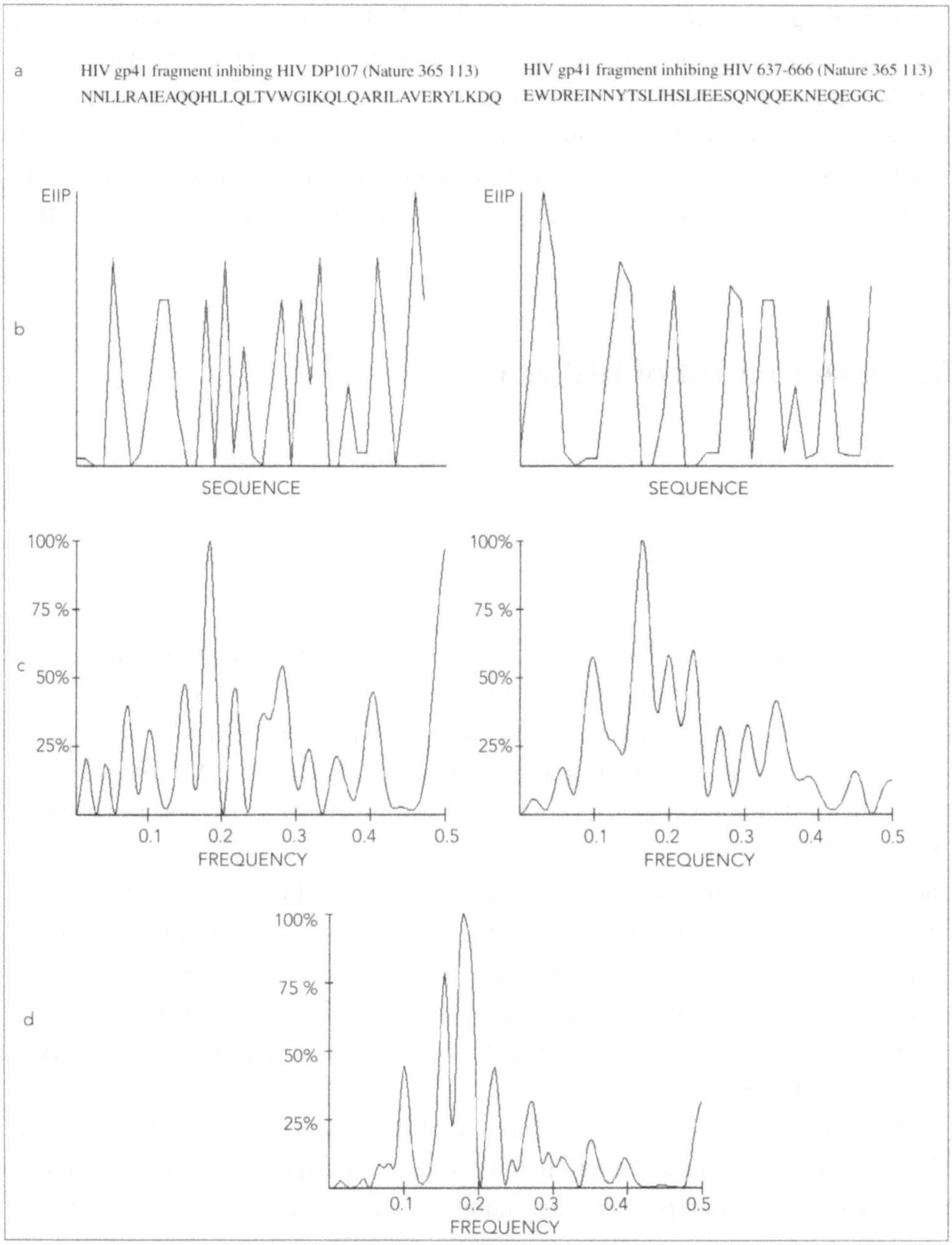

Figure 7.6 Evaluation of the RRM characteristic frequency for HIV inhibitory peptides

(a) Primary structure of the peptides; (b) corresponding numerical sequence; (c) single spectra of both peptides; (d) cross-spectral function of both peptides. The frequency component at 0.185 is prominent in both single spectra as well as in the cross-spectral function.

According to assumptions underlying the RRM (same characteristic frequency and opposite phase), there are sufficient conditions for the inhibitory peptide DP-107 to interact with gp41 and for peptide 637-666 to interact with gp120. Thus it can be suggested that these peptides inhibit HIV-1 infection by binding either to gp41 or to gp120, preventing gp41-gp120 interaction and consequently HIV fusion with cells. More generally, this example represents additional support to the RRM.

7.6 New concept of HIV activity

The application of the RRM to the human immunodeficiency virus has led to new concepts of HIV behaviour, its infectivity and life cycle. These speculations could produce some answers to the questions arising from the very special behaviour of this virus.

A number of experiments have shown that activation of acquired immunodeficiency syndrome (AIDS) is not only by HIV infection of T4 cells but also through several immunogenic and autoimmunogenic effects [101]. The RRM was employed to identify the parameters that characterise immunoreactivity of HIV proteins. Using these obtained characteristics, peptides were designed *de novo* as described above. Although these peptides do not share any sequence homology with HIV proteins, they were able to generate antibodies against different HIV proteins (p18, p40, p55 and gp120). These results showed that the characteristic relevant to HIV proteins immunoreactivity was identified. The same characteristic was found for the human CD4+ antigen, suggesting that antibodies derived against HIV proteins are able to recognise the CD4+ molecule and are able to cause autoimmune destruction of lymphocytes in AIDS.

"The idea that HIV is the direct cause of AIDS just by infecting T4+ cells and thus is responsible for the depletion of these cells is currently being questioned for many reasons: (a) a surprisingly low number of T cells are infected in AIDS; (b) immunosuppression occurs prior to depletion of CD4 cells; (c) immunity against HIV does not protect against AIDS, i.e. AIDS patients have high quantities of anti-HIV antibodies in their sera and generate a strong anti-HIV cytotoxic T-cell response but are not protected" [101], in fact, people tend to become seriously ill only after they make anti-HIV antibodies; (d) antibodies against HIV proteins can be generated only by introduction of foreign lymphocytes without any contact with the virus [102]. A possible explanation of such behaviour could be that HIV induces a deleterious immune response that attacks the immune system itself [101]. In particular, it has been suggested that the most infected cells in the resting state are "invisible" to the immune system, while many noninfected cells are destroyed by the immune system because they are coated with viral proteins [103] or an immune reaction is triggered by viral proteins that produced some autoimmune effect on T4 lymphocytes [101]. This immune response should be triggered by envelope proteins since they are on the viral surface. Thus, despite the high sequence variability between gp160^{env} from 19 different isolates of HIV, it is expected that at least one specific characteristic needs to be common for all different isolates. This specific common characteristic was identified using the RRM method. One of the frequencies characteristic for HIV envelope-CD4 interaction was identified as the frequency $f_2 = 0.218$. In addition, it has been shown that synthetic peptides which are nonhomologous with any HIV protein but have only the same characteristic frequency as HIV envelope proteins do generate the immune response that can recognise viral proteins. From this result it is possible to propose that the frequency $f_2 = 0.2188 \pm 0.022$ characterises the immune reaction caused by the presence of HIV and its proteins *in vivo*.

However, the same frequency component was found to be the most prominent in the spectra of human CD4+ molecules (Fig. 7.7). This result can provide an explanation for experiments showing the ability of foreign lymphocytes to induce antibodies against HIV proteins without any contact with the virus. Foreign lymphocytes with CD4+ molecules on their surface can induce an immune response to CD4, and in particular at its characteristic frequency f_2. As f_2 characterises HIV envelope proteins, antibodies raised against CD4 (frequency f_2) can recognise HIV proteins. This process is analogous to the results with peptide D presented above.

In addition, since frequency f_2 characterises HIV envelope proteins as well as the CD4 molecule, this frequency could be proposed as characteristic of their mutual recognition. Furthermore, antibodies derived by HIV proteins, or any protein having f_2 as its characteristic, e.g. D peptide or CD4 from foreign lymphocytes, recognise proteins having an f_2 component in their spectra and, consequently, are able to recognise CD4+ and to target and destroy CD4+ cells as well. Thus, we suggest here that the main cause of AIDS is not only infection of T4 cells by HIV but also an autoimmune reaction due to antibodies derived from the presence of HIV proteins *in vivo*. In mice, this effect is expected to be of a lower intensity than in humans, because mouse CD4+ has a less prominent peak at frequency f_2 than it is the case of human CD4+. This might explain AIDS species specificity.

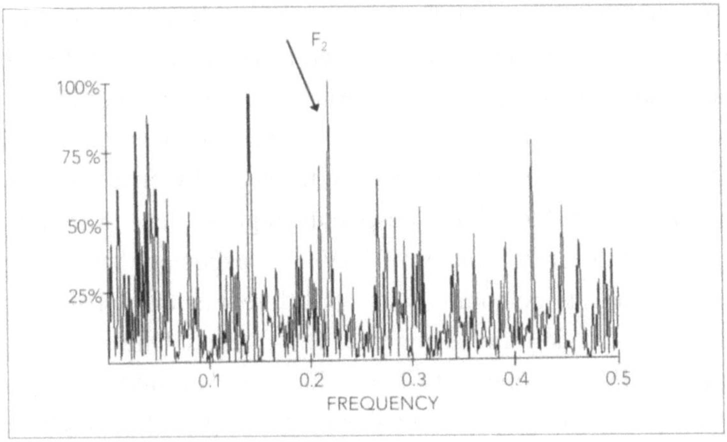

Figure 7.7 The RRM spectrum of the human CD4 molecule

8 Conclusions

The real voyage of discovery consists not in seeking new landscapes but in having new eyes. – Proust

8.1 New concept in molecular biology, new point of view

This book presents a completely new and uncommon concept of selective interactions in linear macromolecules such as proteins, DNA and RNA relevant to their biological function. The physical nature of the biological function of a protein or DNA is based on the ability of the macromolecule to selectively interact with particular targets (other proteins, DNA regulatory segments or small molecules). These interactive processes are the basis of living processes within cells and tissues, and each involves molecular interactions and consequently energy transfer between interacting molecules. Selectivity is defined within the protein structure, but the physical nature of these interactions is not yet well understood. The most widely accepted model is the so-called lock-and-key model, which attributes selectivity to spatial complementarity between interacting molecules. But a thorough examination of a greater number of 3D structures of proteins and their complexes with ligands reveals that spatial complementarity is not selective enough to be the sole parameter controlling protein interactions.

We have discussed selectivity in a completely new light. As all selective processes are resonant in nature, we have proposed that macromolecular interactions represent the transfer of resonant energy between interacting molecules. We have further predicted that these energies are electromagnetic in nature. The characteristic resonant frequencies for a number of different interactions, i.e. biological functions, were theoretically calculated. based on the following key finding: Proteins with the same biological functions do have common periodicities in the distribution of the energies of delocalised electrons along the protein.

With this in mind, and taking into account the conductive properties of the protein backbone, a theoretical model of biologically relevant protein resonant frequencies was established. These frequencies were calculated and found to cover a very wide frequency range including the infrared and visible light.

 ©1997, Birkhäuser Verlag Basel

In order to justify these propositions, calculated resonant frequencies for particular protein functions were compared with the characteristic frequencies of light-absorbing and light- activating proteins. Both calculated and measured frequency characteristics matched each other very closely. The model presented generally describes a new perspective of biologically relevant inter-molecular interactions which are suggested to consist of resonant energy transfer between interacting molecules. Although it is now possible to calculate characteristic frequencies for particular interactions or biological functions, the exact physical basis of these interactions is still not well understood.

With the knowledge of characteristic frequencies for particular interactions (functions), it is possible to predict active sites, functional mutations and even to design bioactive peptides with desired functions. This could have a great impact on molecular biology and consequently on medicine, pharmacology and agriculture. The fact that completely new, nonhomologous but biologically active peptides were designed solely from a knowledge of characteristic frequencies is further evidence that these theoretically calculated frequencies are in fact the critical parameter for protein biological functions, i.e. selective interactions.

8.2 Advances in theory: macromolecular physics

It is an obvious fact that sunlight is the main cause and support of life. It was once claimed that heat from the sun and sunlight are the main catalyst for the transformation of inorganic substances into organic ones [60]. However, macromolecules, mainly proteins, through their highly selective interactions with different targets drive most of the processes of life. If the first organic molecules performed their biological function within the energy range of sunlight at the dawn of life, we can expect that more complicated organic molecules, though with the same biological functions, will maintain their functioning within the same energy (frequency) range.

Sunlight → origin of life ↘
Proteins interact using sunlight energy
Protein interactions → life processes ↗

All the results obtained so far with the RRM lead towards the conclusion that each macromolecular biological process inside the living cell or tissue is characterised by one frequency within a very wide frequency range from extremely low infra-red up to ultraviolet. Heat from the sun and sunlight are electromagnetic radiation emitted in a range of frequencies from the extremely low infrared to ultraviolet, the same range of frequencies predicted by the RRM to be responsible for protein-protein and protein-DNA interactions, i.e. the main intermolecular processes of life. Does this mean that life's intermolecular processes are carried out at their own frequencies and these frequencies lie within sun light frequency range?

8.3 Advances in molecular biology; design of new pharmaceuticals

The concepts presented here have led to a completely new perspective on life processes at the molecular level. Consequently, they represent a step towards a better understanding of biological processes and biomacromolecular interactions. The RRM is capable of bringing a number of practical advantages to the fields of molecular biology, biotechnology, medicine and agriculture. It is now possible:

- to design *de novo* a variety of polypeptide analogues which will have desired RRM spectral characteristics and consequently desired functions;
- to identify mutations in a protein that can modify the protein function in a de sired manner;
- to identify fragments of the original protein that will have desired spectral characteristics and functions.

For example, the RRM could be used to identify protein active sites as described in Chapters 4 and 5. This will enable scientists to predict functional mutations in proteins and DNA and can conceivably lead towards rationally modified drugs or other biochemicals.

Rational *de novo* design of a variety of peptides and proteins would have an enormous impact on future biotechnology, medicine and the pharmacology industry. With a new method to design drugs, pharmaceuticals and biotechnology products with desired biological functions.

In terms of molecular biology, the RRM could have a substantial impact on the engineering of DNA, the design of more potent promoters or enhancers for insertion in plasmids or the identification of functional DNA segments. However, thus far only a few products designed with the RRM have been tested experimentally. Results are promising but need to be improved. An exciting adventure awaits us in accepting, improving and testing the RRM as well as in utilising it for an impressive range of applications.

References

1 Dayhoff, M.O. (1978) *Atlas of Protein Sequence and Structures* vol. 5, suppl. 3, pp. 345-352, Natl. Biomed. Res. Fund. Washington, DC.

2 Devereux, J., Haeberly, P., Smithies, O. (1984) A comprehensive set of sequence analysis programs for the VAX, *Nucl. Acids Res.* **12**: 387-395.

3 Doolittle, R. (1981) Similar amino acid sequences: chance or common ancestry? *Science* **214**: 149-159.

4 Breg, J.W., van Opheusden, J.H.J., Burgering, M.J.M., Captein, R. (1990) Structure of Arc repressor in solution: evidence for a family of β-sheet DNA-binding proteins, *Nature* **346**: 586-589.

5 King, (1989) Deciphering the rules of protein folding, *Chem. Eng. News* **32**.

6 Fasman, G.D. (ed.) (1989) *Prediction of Protein Structure and the Principles of Protein Conformation* Plenum Press, New York.

7 Chou, Y., Fasman, G.D. (1975) The conformation of glucagon: predictions and consequences, *Biochem.* **14**: 2536-2540.

8 DeLisi, C., Berzofsky, J.A. (1985) T-cell antigenic sites tend to be amphipatic structures, *Proc. Natl. Acad. Sci. USA* **82**: 7048-7052.

9 Schmid, E.D., Schneider, F.W., Siebert, F. (1988) *Spectroscopy of Biological Molecules: New Advances* John Wiley & Sons, New York.

10 Mackay, D.H.J., Cross, A.J., Hagler, A.T. (1989) The role of energy minimisation in simulation strategies of biomolecular systems, *in:* G. D. Fasman (ed.): *Prediction of Protein Structure and the Principles of Protein Conformation* pp. 317-357, Plenum Press, New York.

11 Maple, J., Dinur U., Hagler, A.T. (1988) Derivation of force field for molecular mechanics and dynamics from ab initio energy surfaces, *Proc. Natl. Acad. USA* **85**, 5350-5354.

12 Karplus, M., McCammon, J.A. (1983) Dynamics of proteins: elements and function, *Ann. Rev. Biochem.* **53**: 263-300.

13 Rosenberg, M., Court, D. (1979) regulatory sequences involved in the promotion and termination of RNA transcription, *Ann. Rev. Genet.* **13**: 319-353.

14 Benerji, J., Olson, L., Schaffner, W. (1983) Expression of a β globine gene is enhanced by remote SV40 DNA sequence, *Cell* **33**: 729.

15 Rost, B., Schneider, R., Sander, C. (1993) Progress in protein structure prediction, *Trends in Biochemical Science (TIBS)* **18**: 120-123.

16 Ladik J. (1974) All valence electron band structures of simple periodic protein models, *Int. J. Quantum Chemistry Quantum Biol. Symp* **1**: 65-69.

17 Little, W.A. (1964) Possibility of synthesizing an organic superconductor, *Phys. Rev.* **134**: A1416-1424.

18 Davidov, A.S. (1981) The role of solitons in the energy and electron transfer in one dimensional molecular systems, *Physica 3D–Non Linear Phenomena* **3**: 1-22.

19 Davidov, A.S. (1982) *Biology and Quantum Mechanics* Pergamon Press, Oxford.

20 Jensen, P., Jaric, M., Bennemann, K.H.

 ©1997, Birkhäuser Verlag Basel

(1983) Soliton like processes during right-left transition in DNA, *Phys. Lett.* **95A**: 204-208.

21 Kendrew, J.C., Bodo, G., Dinatzis, H.M., Parrish, R.G., Wyckoff, H., et al. (1958) A three-dimensional model of the myoglobin molecule obtained by X-ray analysis, *Nature* **81**: 662-666.

22 Skolnik, J. (1989) Computer simulations of globular protein folding and tertiary structure, *Ann. Rev. Phys.* Chem. **40**: 207-235.

23 Scheraga, H.A. (1973) *Current Topics in Biochemistry* pp. In: 1-42, C.B. Anfinsen, Schechter (eds) A.N., Academic, New York.

24 Cosic, I. (1990) Resonant recognition model of protein-protein and protein-DNA interaction, In: D. Wise (ed.) *Bioinstrumentation and Biosensors* pp. 475-510, Marcel Dekker, New York.

25 Cosic, I. (1994) Macromolecular bioactivity: is it resonant interaction between macromolecules? Theory and Applications, *IEEE Trans. on Biomedical Engineering* **41**: 1101-1114.

26 Cosic, I. (1995) Virtual spectroscopy for fun and profit, *Bio/Technology* **13**: 236-238.

27 Cosic, I., Nesic, D., Pavlovic, M., Williams, R. (1986) Enhancer binding proteins predicted by Informational Spectrum Method, *Biophys. Biochem. Res. Comm.* **141**: 831-839.

28 Cosic, I., Nesic, D. (1988) Prediction of "hot spots" in SV40 enhancer and relation with experimental data, *Eur. J. Biochem.* **170**: 247-252.

29 Cosic, I., Pavlovic, M., Vojisavljevic, V. (1989) Prediction of "hot spots" in interleukin-2 based on informational spectrum characteristics of growth regulating factors, *Biochimie* **71**: 333-342.

30 Cosic, I., Vojisavljevic, V., Pavlovic, M. (1989) The relationship of the resonant recognition model to effects of low-intensity light on cell growth, *Int. J. Radiat. Biol.*, 56: 179-191.

31 Cosic, I., Hodder, A., Aguilar, M., Hearn, M.T.W., (1991) Resonant recognition model and protein topography: model studies with myoglobin, hemoglobin and lysozyme, *Eur. J. Biochem.* **198**: 113-119.

32 Cosic, I., Hearn, M.T.W. (1991) "Hot spot" amino acid distribution in ha-*ras* oncogene product p21: relationship to guanine binding site, *J. Mol. Recognition* **4**: 57-62.

33 Cosic, I., Hearn, M.T.W. (1992) Studies on protein-DNA interactions using the resonant recognition model: application to repressors and transforming proteins, *Eur. J. Biochem.* **205**: 613-619.

34 Cosic, I., Drummond, A.E., Underwood, J.R., Hearn, M.T.W. (1993) A new approach to growth factor analogue design: modelling of FGF analogues, *Mol. Cell. Biochem.* **130**: 1-9.

35 Cosic, I. (1988) Protein design by methods of digital signal analysis, *In: Proceedings of Annual Conference IEEE BME*, vol. 10, pp. 901-905, IEEE Publ. Service, New York.

36 Kanehisa, M. (1988) A multivariate analysis method for discriminating protein secondary structural segments, *Protein Eng.* **2**: 87-92.

37 Nakai, K., Kidera, A., Kanehisa, M. (1988) Cluster analysis of amino acid indices for prediction of protein structure and function, Protein Eng. 2: 93-100.

38 Veljkovic, V. (1980) *A Theoretical Approach to Preselection of Cancerogens and Chemical Carcinogenesis*, Gordon & Breach, New York.

39 Charton, M. (1981) Electrical effect substituent constants for correlation analysis, *Prog. Phys. Organic Chem.* **13**: 119-251.

40 Charton, M., Charton B.I. (1982) The structural dependence of amino acid hydrophobicity parameters, *J. Theor. Biol.* **99**: 629-644.

41 Heine,V., Koen, M., Uir, D. (1973) *Theory of Pseudopotentials*, MIR, Moscow.

42 Veljkovic, V., Slavic, I. (1972) General model of pseudopotentials, *Physical Review Lett.* **29**: 105-108.
43. Rabiner, L., Gold B. (1975) *Theory and Application of Digital Signal Processing*, Prentice Hall, New York.
44. Mandell, A. (1988) *Chaos and order in the natural science*, Springer-Verlag, Berlin.
45. Viari, A., Soldano, H., Ollivier, E. (1990) A scale-independent signal processing method for sequence analysis, Comput. Appl. *Biosci.* **6**: 71-80.
46. Cosic, I. (1994) Bioactive peptide design using the Resonant Recognition Model: application on nerve growth factors, *Miami Bio/Technology* **1**: 29.
47. Cosic, I., Fang, Q. (1995) Problems in using FFT versus DFT in the Resonant Recognition Model, *Proc. of IEEE EMBS* **17**: 289-290.
48. Fang, Q., Cosic, I. (1996) *Refinement of the Resonant Recognition Model using digital windows and Blackman-Tukey method*, Technical Report TR-96-9 Monash University.
49. Tukey, J.W., Blackman, R.B. (1959) The *Measurement of Power Spectra from the Point of View of Communications Engineering*, Dover, New York.
50. Blum, H. (1959) *Carcinogenesis by Ultraviolet Light*, Princeton University Press, Princeton, NJ.
51. Williams J., Fox, L. (1980) Coupled oscillator theory of enzyme action, *In: Chemical Recognition in biology*, pp. 86-96, Springer-Verlag.
52. Halhaus, J., Thomas, D., Hagness, D. (1986) Molecular genetics of human color vision: the genes encoding blue, green and red pigments, *Science* **232**: 193-201.
53. Inouye, S., Noguchi M., Sakaki, Y., Takagi, Y., Miyata, T., Iwanaga, S., Miyata, T., Tsuji, F. (1985) Cloning and sequence analysis of cDNA for the luminescent protein aequorin, *Proc. Natl. Acad. Sci. USA* **82**: 3154-3158.
54. Karu, T.J. (1987) Photobiological fundamentals of low-power laser therapy, *IEEE Journal of Quantum Electronics* **QE-23**: 1703-1717.
55. Biscar, J.P. (1976) Photon enzyme activation, *Bull. Math. Biol.* **38**: 29-38.
56. Cosic, I. (1994) Correlation between predicted and measured characteristic frequency of chymotrypsin activation, *Proc. IEEE EMBS* **15**: 265-266.
57. Liscum, E., Hangarter, R.P. (1991) Arabidopsis mutants lacking blue light dependent inhibition of hypocotyl elongation, *Plant Cell* **3**: 685-694.
58. Ahma, M., Cashmore, A.R. (1993) HY4 Gene of *A. Thaliana* encodes a protein with characteristics of blue-light photoreceptor, *Nature* **366**: 162-166.
59. Cosic, I., Birch, S. (1994) photoreceptors having similar structure but different absorptions can be distinguished using the Resonant Recognition Model, *Proc. IEEE EMBS* **16**: 265-266.
60. Tesla, N. (1915) How cosmic forces shape our destinies?, *New York American*, **7 Feb., A**: 172-174.
61. Kirshenbaum, D. (ed.) (1974) Atlas of Protein Spectra in the Ultraviolet and Visible Region, IFI/Plenum Press, New York.
62. Frolich, H. (1985) Further evidence for coherent excitations in biological systems, *Phys. Lett.*, **110A**, 480-481.
63. Lerner, E. (1984) *Biological effects of electromagnetic fields*, IEEE Spectrum, May, 57-67.
64. Veljkovic, V., Cosic, I. (1987) A novel method of protein analysis for prediction of biological function: application to tumor toxins, *Canc. Biochem. Biophys.* **9**: 139-148.
65. Brookhaven Protein Data Base, 1995.
66. Nicholls, P. (1975) *Cytochromes and Biological Oxidation* Oxford, London.
67. Zenke, M., Grundstrom, T., Matthes, H., Wintzerith, M., Schatz, C., Wildeman, A., Chambon, P. (1986) Multiple sequence motifs are involved in SV40 enhancer function, *EMBO J.* **5**, 387-397.

68 Wildeman, A., Zenke, M., Schatz, C., Takahashi, K., Barrera-Saldana, H., Grundstrom, T., Wintzerith, H., Matthes, H., Vigneron, M., Chambon, P. (1985) *Current communication in molecular biology: eucariotic transcription*, pp. 19-27, Cold Spring Harbor Laboratory, Cold Spring Harbor, New York.

69 Cohen, E.E., Kosen, P. A., Kuntz, I. D.,Epstein, L. B. , Ciardelli, T. L., Smith, K.A., (1986) *Structure-activity studies of interleukin-2, Science*, **234**: 349-352.

70 Cosic, I., Pavlovic, M. (1986) Informational spectrum method applied to growth factors, pp. 309-312, *Proc. Med. Biol. Eng.*, Puntex, Barcelona.

71 Mroczkowski, B., Ball, R. (1990), Epidermal growth factor: biology and properties of its gene and protein precursor, *In:* A. Habenicht (ed.) *Growth factors, differentation factors and cytokines*, Springer-Verlag, Berlin.

72 Bode, W. (1979) The transition of bovine trypsinogen to a trypsin-like state upon strong ligand binding, *J. Mol. Biol.* **127**: 357.

73 Appel, W. (1986) Chymotrypsin: molecular and catalytic properties, *Clinical Biochem.* **19**: 317-322.

74 Marquart, M., Walter, J., Deisenhofer, J., Bode, W., Hyber, R. (1983) The geometry of the reactive site and the peptide groups in trypsin, trypsinogen and its complexes with inhibitors, *Acta. Cryst.* **B39**: 480-490.

75 Kahn, P., Graf, T. (1986) Oncogenes and Growth Control, Springer-Verlag, Berlin.

76 Pai, E.F., Kabsch, W., Krengel, U., Holmes, K.C., John, J., Wittinghofer, A. (1989) Structure of the guanine-nucleotide-binding domain of the Ha-ras oncogene product p21 in the triphosphate conformation, *Nature*, **341**: 209-214.

77 Barbacid, M. (1987) Ras genes, *Ann. Rev. Biochem.* **56**: 779-827.

78 Veljkovic, V., Metlas, R., Raspopovic, J., Pongor, S. (1992) Spectral and sequence similarity between vasoactive intestinal peptide and the second conserved region of human immunodeficiency virus type 1 envelope glycoprotein (gp120): possible consequences on prevention and therapy of AIDS, *Biochem. Biophys. Res. Commun.* **189**: 705-710.

79 Rechler, M.M. (1993) Insulin-like growth factor binding proteins, *Vitamins and Hormones* **47**: 1-114.

80 Randen, I., Potter, K.N., Li, Y., Thompson, K.T., Pascual, V., Forre, O., Natvig, J.B., Capra, J.D. (1993) Complementarity-determiniing region 2 is implicated in the binding of staphylococcal protein A to human immunoglobulin V_{HIII} variable regions, *Eur. J. Immunol.* **23**: 2682-2686.

81 Seppala, I., Kaartinen, M., Ibrahim, S., Makela, O. (1990) Mouse Ig coded by V_H families S107 or J606 bind to protein A, *Int. J. of Immunol.* **145**: 2989-2993.

82 Schaffner, W. (1985) *Current communications in molecular biology: eucaryotic transcription*, Cold Spring Harbor Laboratories, Cold Spring Harbor, NY.

83 Ptashne, M. (1986) *A Genetic Switch*, Cell Press & Blackwell Scientific Publishers, Cambridge.

84 Schevitz, R., Otwinowski, Z., Joachimiak, A., Lawson, C.L., Sigler, P.B. (1985) The three-dimensional structure of trp repressor, *Nature* **317**: 782-786.

85 Luisi, B.F., Xu, W.X., Otwinowski, Z., Freedman, L.P., Yamamoto, K.R., Sigler, P.B. (1991) Crystallographic analysis of the interaction of the glucocorticoid receptor with DNA, *Nature* **352**: 497-505.

86 Bollag, G., McCormic, F. (1991) Differential regulation of rasGap and neurofibromatosis gene product activities, *Nature* **351**: 576-579.

87 Owen, R.D., Bortner, D.M., Ostrowski, M.C. (1990) Ras oncogene activation of a VL30 transcriptional element is linked to transformation, *Mol. Cell. Biol.* **10**: 1-9.

88 Ptashne, M., Gann, A.A.F. (1990) Activators and targets, *Nature* **346**: 329-331.

89 Sassone-Corsi, P., Der, C.J., Verma, I.M. (1989) Ras-induced neuronal differentation of PC12 cells: possible involvement of fos and jun, *Mol. Cell. Biol.* **9**: 3174-3183.

90 Crouch, M.F., Belford, D.A., Milburn, P.J., Hendry, I.A. (1990) Pertussis toxin inhibits EGF-, phorbol ester-, and insulin-stimulated DNA synthesis in BALB/c3T3 cells: evidence for post-receptor activation of Giα, *Biochem. Biophys. Res. Commun.* **167**: 1369-1376.

91 Kovary, K., Armelin, M.C., Armelin, H.A. (1989) Glucocorticoid dexametasone reversibly complements EJ-RAS oncogene to transform mouse embryo BALB-3T3 cells, *J. Cell Bioch* **41**: 171-177.

92 Hruby, V.J. (1982) Structure-conformation-activity studies of glucagon and semi-synthetic glucagon analogues, *Mol. Cell Biochem.* **44**: 49-64.

93 Krstenasky, J.L., Zechel, C., Trivedi, D., Hruby, V.J. (1988) Importance of the C-terminal -helical structure for glucagon's biological activity, *Int. J. Peptide Protein Res.* **32**: 468-475.

94 Unson, C.G., Gurzendia, E.M., Iwasak, K., Merrifield, R.B. (1989) Glucagon antagonists: contribution to binding and activity of the amino-terminal sequence 1-5, position 12, and the putative α-helical segment 19-27, J. Biol. Chem. **264**: 789-794.

95 Korn, P., Ottensmeyer, F.P. (1983) A model for the three-dimensional structure of glucagon, *J. Theoret. Biol.* **105**: 403-425.

9 Krsmanovic, B., Cosic, I., Biquard, J.M., Hearn, M.T.W. (1993) *Artificial peptides and induced antibodies*, Patent Appl. PCT/FR 93/00171.

97 Goustin, A.S., Leof, E.B., Shipley, G.D., Moses, H.L. (1986) Growth factors and cancer, *Cancer Res.* **46**: 1015-1029.

98 Anderson, R. (1993) AIDS: trends, predictions, controversy, *Nature* **363**: 393-394.

99 Lasky, L.A., Nakamura, G., Smith, D.A., Fennie, C., Shimasaki, C., Patzer, E., Berman, P., Gregory, T., Capon, D.J. (1987) Delineation of a region of the human immunodeficiency virus type 1 gp120 glycoprotein critical for interaction with the CD4 receptor, *Cell* **50**: 975-985.

100 Jiang, S., Lin K. Strick N., Heurath A. R., (1993) HIV-1 inhibition by a peptide, *Nature* **365**: 113.

101 Hoffmann, G.W., Kion, T.A., Grant, M.D. (1991) An idiotypic network model of AIDS immunogenesis, *Proc. Natl. Acad. Sci. USA* **88**: 3060-3064.

102 Kion, T.A., Hoffmann, G.W. (1991) Anti-HIV and anti-anti-MHC antibodies in alloimmune and autoimmune mice, *Science* **253**: 1138-1140.

103 Klatzmann, D., Montagnier, L. (1986) Approaches to AIDS therapy, *Nature* **319**: 1011.

Index

BIOLOGY • MOLECULAR BIOLOGY

T. Meier, *University of Colorado, Denver, CO, USA*
F. Fahrenholz, *Max-Planck-Institut für Biophysik, Frankfurt/M., Germany (Eds)*

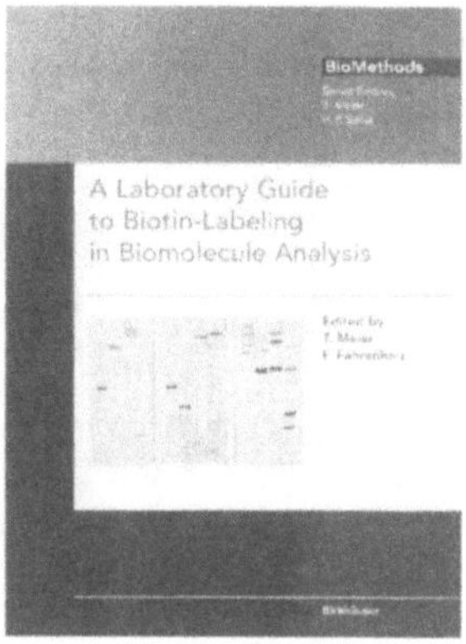

A Laboratory Guide to Biotin-Labeling in Biomolecule Analysis

1996. 256 pages. Hardcover • ISBN 3-7643-5206-X

This book summarizes protocols and applications of recently developed or improved non-radioactive biotin-labeling techniques for proteins, glycoproteins and nucleic acids and will provide valuable help to researchers both in fundamental and in applied sciences.

The first chapter of this volume compares the chemical properties of biotin-labeling compounds currently available and outlines their reaction principles. The following contributions provides a step-by-step protocol on how to prepare and successfully apply biotin-labeled probes for the analysis of complex biochemical and cellular systems. An extended troubleshooting section completes each of the protocols. In most cases these core protocols provide a guideline that encourages modifications according to the researchersí experimental designs.

Combined with sensitive detection, these recently developed experimental procedures are powerful tools for many applications in areas ranging from protein biochemistry to molecular and cellular biology

From the Contents:
Biotinylation and chemical crosslinking of membrane associated molecules • Application of biotinylated lectins in glycobiology research • Photocleavable biotinylated ligands for affinity chromatography • Biotin *in vitro* translation, a nonradioactive method for the synthesis of Biotin-labeled proteins in a cell free system • Immunoprecipitation of biotinylated cell surface proteins • Preparation and use of biotinylated probes for the detection and characterization of serine proteinase and serine proteinase inhibitory proteins • Avidin/Biotin-mediated conjugation of antibodies to erythrocytes: An approach for immunoerythrocytes exploration *in vivo* • Purification of the receptor for pituitary adenylate cyclase-activating polypeptide (PACAP) using biotinylated ligands

Birkhäuser Verlag • Basel • Boston • Berlin

BioMethods

H.P. Saluz, *IRBM, Pomezia (Roma) Italy*
J.P. Jost, *Friedrich-Miescher-Institut, Basel, Switzerland (Eds)*

A Laboratory Guide for *In Vivo* Studies of DNA Methylation and Protein/DNA Interactions

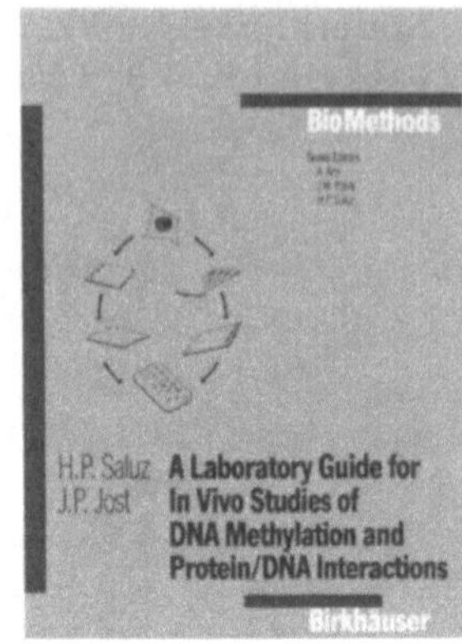

1990. 288 pages. Hardcover • ISBN 3-7643-2369-8
(BioMethods 3)

«Researchers interested in studying DNA modifications, mutations, and protein binding to genes as they are present in the living individual, tissue or cell are greatly helped by this most concise and instructive laboratory guide.»
Diter von Wettstein, PhD Carlsberg Laboratory, Copenhagen, Denmark

«The scientific dream of reading DNA sequences and modifications directly within the cell has become reality with the use of genomic sequencing. The two most successful pioneers in this technology, Drs. H.P. Saluz and J.-P. Jost, provide a comprehensive and exceptionally useful manual covering the theoretical and laboratory aspects of genomic sequencing. Their book is an absolute requisite for those who would either understand or utilize this powerful tool. As such, the book is the only one of its genre in scientific literature.»
Arthur Weissbach, PhD
Roche Institute of Molecular Biology, Nutley, USA

Birkhäuser Verlag • Basel • Boston • Berlin

BioMethods

J.P. Jost, *Friedrich Miescher-Institut, Basel, Switzerland*
H.P. Saluz, *IRBM, Rome, Italy (Eds)*

A Laboratory Guide to *In Vitro* Studies of Protein-DNA Interactions

1991. 328 pages. Hardcover • ISBN 3-7643-2627-1

With contributions by Sankar Adhya, Nigel Grindley, Tim Reudelhuber, Walter Schaffner, Fred Schaufele, Stuart R. Stone, Thomas D. Tullius, Dietmar Von der Ahe, Zhou Wang, Walter Wahli

The aim of this multi-author laboratory guide is to provide a comprehensive collection of the major techniques used for the detailed study of protein/DNA interactions in vitro. It is divided into 21 chapters and 10 appendices devoted to the description of procedures including: in vitro footprinting (involving enzymes, UV-light, free radicals and chemicals), characterization of protein/DNA complexes (by electron microscopy, UV cross-linking, methylation interference and missing contact probing) and detection of unusual DNA structures using „bending" vectors. In addition, several associated techniques are described. For instance, qualitative one-dimensional and two-dimensional gel shift assays as well as the kinetic aspects of protein/DNA interactions are clearly outlined in detail with specific examples. The use of DNA amplification for in vitro footprinting is also included. A section on protein/DNA interactions on solid supports gives protocols for the affinity chromatography of DNA-binding proteins by screening cDNA expression libraries with labeled binding-site probes.

This collection of detailed protocols should prove very helpful for anyone interested in the qualitative and quantitative studies of the interactions between protein factors and DNA.

Sections from the Contents: The Study of Protein-DNA Interactions by Means of Enzymes • The Study of Protein-DNA Interactions by UV Light • The Study of Protein-DNA Interactions by Means of Chemicals (Interference Assays) • The Study of Protein-DNA Interactions by Means of Electron Microscopy • The Study of Protein-DNA Interactions by Gel-Mobility-Shift Assays • Protein-DNA Interactions on a Solid Support • The Study of Protein-DNA Interaction Using Special Vectors • Other Related Techniques • Appendices • Index

Birkhäuser Verlag • Basel • Boston • Berlin

MIX
Papier aus verantwortungsvollen Quellen
Paper from responsible sources
FSC® C105338

If you have any concerns about our products,
you can contact us on
ProductSafety@springernature.com

In case Publisher is established outside the EU,
the EU authorized representative is:
Springer Nature Customer Service Center GmbH
Europaplatz 3, 69115 Heidelberg, Germany

Printed by Libri Plureos GmbH
in Hamburg, Germany